I0493852

Identifying Identical Twin Star Systems from the SDSS Data Release 10

by

Martin P Nicholson

Copyright © 2014 by Martin Nicholson

All rights reserved. No part of this publication may be reproduced, distributed, or transmitted in any form or by any means, including photocopying, recording, or other electronic or mechanical methods, without the prior written permission of the author, except in the case of brief quotations embodied in critical reviews and certain other noncommercial uses permitted by copyright law. For permission requests email the author at the address below.

Martin Nicholson
Church Stretton
Shropshire SY6 7DQ
United Kingdom

Email – newbinaries@yahoo.co.uk

A message from the author.

My New Year resolution for 2014 – "The results of all astronomical projects I have done in the past, or that I will do in the future, must be published and made available to the wider astronomical community."

Looking back, every year, almost without fail, I have found myself standing in silence in memory of another hobby personality who had died. It didn't matter if the deceased was an astronomer, a "grave hunter" or a postal historian - so often the legacy they left was greatly reduced because so much of their knowledge and experience died with them.

Only a small proportion of the astronomical data mining and astronomical imaging projects I carried out over the last 20 years went through the lengthy – and sometimes controversial – process of third-party publication. Further details of these papers can be found on this web page:

http://www.martin-nicholson.info/1/success.htm

Some of the rest of my work has appeared on assorted web sites or within specialist society "news groups" but much of it remains unpublished. The long-established scientific principle "first to publish gets the credit" means that without publication taking place it is almost as if the work had never been done! Confusingly, there seems to be no consensus about what constitutes "publishing" but my 2009 thoughts on the subject still represent my position.

http://www.philica.com/display_article.php?article_id=164

"The vsnet-alert list exists to "To distribute alert notices of important phenomena". It is an un-moderated group but self-regulation works well and the vast majority of postings appear without generating any adverse comment. Sadly there are a few users of the facility who routinely post off-topic material. These postings vary in nature from the harmless posting of scientifically valid astronomical material that would perhaps have been better posted elsewhere right through to lengthy personal attacks with not a vestige of scientific reasoning to be seen.

In July 2009 vsnet-alert 11306 was posted by "John". Although hopelessly off-topic it did raise some interesting issues regarding what constitutes publication.

It is not in dispute that the article to which "John" referred - "Identifying Previously Uncatalogued Red Variable Stars in the Northern Sky Variability Survey" - is available to the public and that the relevant community has been made aware of its existence.

The American Association of Variable Star Observers (AAVSO) website also clearly states that, "The eJAAVSO consists of papers that have been refereed, edited, and accepted for publication in the paper edition of the JAAVSO." This means that long term access is in place, particularly so once paper-based publication has taken place.

"John" makes some, doubtless "tongue in cheek", quotes from the article "I've submitted all this that and the other to AAVSO VSX and this is a note saying so, and here are some light curves of about half a dozen of the bestest (sic) ones so all the unillustrated 1200+ L: must be as good too honest, therefore I've now published these stars and they are mine despite my not having actually published a table or light curves (just a link to a spreadsheet somewhere) nor done any further checks of the literature for them since January 2007".

I imagine the point that "John" is trying to make here is that when 1,233 new discoveries need to be reported what constitutes publication? It is unlikely that any publication would be prepared to publish 1,233 individual light curves. Broadly similar papers published by the Open European Journal on Variable Stars, such as OEJV #105, show the issues associated with showing more than perhaps six or at a maximum eight light curves per page. Publishing 1,233 light curves would require over 150 pages! The adopted solution - publishing a selection with links to the remainder - was the preferred solution of the author, the professional referee and the journal editor.

The final sentence of vsnet-alert 11306 reads, "But the above lot will now be claimed to be published in a peer reviewed journal, you watch and see." "John" is quite correct in this view - the results were published using any standard definition of the word - and the peer reviewed status of eJAAVSO is well established."

"Research Topics for Amateur Astronomers" will be a multi-volume series containing two types of material. *Research Notes* are intended to share discoveries, ideas or techniques of interest to astronomers. *Articles* equate to a traditional peer reviewed article.

Martin Nicholson – Shropshire, UK.
February 2014

3

Identifying "Identical Twin" Star Systems from the SDSS Data Release 10

Martin P. Nicholson

Ticklerton Barn, Ticklerton, Church Stretton,
Shropshire, England, SY6 7DQ

e-mail: newbinaries@yahoo.co.uk

Abstract: Data mining the SDSS Data Release 10 has yielded a total of 481 pairs of stars separated by between 3 and 30 arc second where all five listed de-reddened magnitudes (u, g, r, i and z bands) for the two stars are within 0.05 magnitudes. 473 of these binary stars are believed to be new discoveries.

Introduction: Astronomical data mining, particularly in the fields of binary star identification and characterisation, is one of the few branches of science where amateurs can still make significant observations or discoveries.

Two stars orbiting around their joint centre of mass are called a binary star system. Binary stars are important in astrophysics for a number of reasons - not least because orbital studies allow the mass of the stars to be determined. Various sub-types of binary stars exist such as optical binaries, spectroscopic binaries and eclipsing binaries.

Optical double stars are just "line of sight" arrangements of no scientific importance.

Historically it was very difficult to distinguish between the wheat (binary stars) and the chaff (double stars). For this reason the Washington Double Star Catalog (WDS) is comprised of a mixture of three types of object.

- Binary stars
- Double stars
- Stars that, without further information, might fall into either class

For many years just three factors had to be reported when discussing an observation of a binary star – the position in the sky, the separation and the position angle.

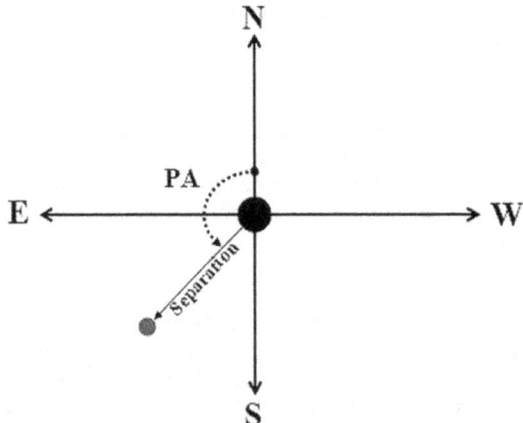

Figure 1 – The key features of a double or binary star system

Most of the pairs observed by amateur astronomers change very slowly and frequent measurements were unnecessary. This encouraged a switch from <u>measuring</u> binary stars to searching for <u>previously unreported</u> examples.

In July 2002 I contacted the United States Naval Observatory seeking advice as to what would qualify as a new pair worth reporting.

One section of the reply I received from Dr. William I. Hartkopf (July 2002) of the Astrometry Department is worth quoting in full. (Journal of the British Astronomical Association, Vol. 115, No. 6, p.342)

"As for qualifications - what pairs are worth reporting - that's a more difficult question! The WDS contains a rather eclectic mix of physical and optical pairs - essentially any published double. Some are very wide and are only included because someone thought they were, for example, a common-proper motion pair, etc. and published a measurement! (In other words, they're included for historical interest or "completeness", but are otherwise of little use.) Physical pairs are most interesting from an astronomical standpoint, but it's very difficult to determine with any degree of certainty whether a pair is indeed physical. Several "rules of thumb" have been published, however. Robert Aitken used the following formula in one of his books: log p = 2.8 - 0.2 m, where p is the separation in arc-seconds and m is the apparent visual magnitude (probably combined magnitude, although it isn't mentioned). Binaries of a given magnitude closer than p are considered probably physical. Aitken (and others who came up with similar formula)would often include "borderline" cases if they found them interesting, so these rules weren't set in stone! For your purposes you might use this formula as a guideline, but also keep any systems fainter than 9th magnitude if their separations are 10" or less."

Brian D. Mason, Project Manager, Washington Double Star Program, subsequently contacted me (December 2002) saying:

"What criteria did you use for identifying new doubles? When we have gone through astrometric catalogs (like the AC) looking for new doubles we've only selected ones meeting one of two possible parameter: a. Aitken's parameter : described in the introduction to the ADS, this relates apparent magnitude and separation, b. outer limit : we arbitrarily set this at ten arc seconds, i.e., if the two stars are at 10" and closer we call it a double and put it in the WDS, even if it is so faint that the likely physical separation probably rules out it being a real double. If you use criteria similar to this, it might reduce the number of doubles and increase the wheat to chaff ratio."

It soon became clear that the 10 arc second framework *when used on its own* was capable of generating vast numbers of candidate <u>double stars</u> without clearly identifying any <u>binary stars</u>. Indeed it has been claimed that the sheer number of candidate double stars notified to the Washington Observatory in the first few years of the 21st century made the professional astronomers realise that they had set the criteria for inclusion in "their" catalogue incorrectly! It is unfortunate that the subsequent change of policy was not transmitted to researchers in a timely manner.

Luckily a new tool was becoming available and researchers – amateur and professional alike – turned to proper motion as a diagnostic tool.

Careful measurement over many years reveals that all stars are moving independently through space and this creates slow changes in their position relative to the earth. Proper Motion is a vector, which has both a magnitude and a direction. The magnitude has units of arc second per year and the direction is expressed in degrees with 0 degrees being north, 90 degrees being east and so on. Most catalogues present this information in the form of the magnitude of the motion in both right ascension and in declination as these are at right angles to each other.

For a pair to be considered a binary pair the two components would be expected to show very similar proper motion. Reliable and up-to-date results are not available for all stars but where such information is available it is clear that some systems previously reported as binaries were in fact "just" double stars.

One complicating factor was the lack of professional guidance on exactly what constituted "very similar proper motion". This was an unfortunate oversight and led to some over optimistic claims about the binary nature of some pairs. In an article I wrote in 2009 I described in some detail the problems this created.

http://www.philica.com/display_observation.php?observation_id=55

By 2013 most of the stronger candidates for being "common proper motion pairs" had been published – many in the Journal of Double Star Observations (http://www.jdso.org/). It was time to explore the new technique of photometric parallax.

The SDSS provides photometry in five different wavebands and these results can be used to estimate the spectral energy distribution of the star. It is possible to extrapolate from the SDSS results obtained from previously studied stars to the absolute magnitude, and hence the distance, of previously unstudied stars. An extension of this idea of photometric parallax is the idea, presented for the first time in this paper, that if two stars have, essentially, identical re-reddened photometry in all five SDSS wavebands then they must be at the same distance, especially if the angular separation between them is less than 30 arc seconds.

Methodology and Results: The CasJobs facility can be accessed at:

http://skyserver.sdss3.org/CasJobs/

The SQL programme designed to identify "identical twin" star systems is in five parts.

The first section identifies what data fields are to be extracted from the different tables that are available. The "Neighbors" tables contains information on the SDSS objects within 30 arc second of each other and the "PhotoObjAll" table is a full photometric catalogue for each SDSS object.

The second section confines the search to primary objects (mode=1), that are stars (type=6), that have clean photometry (clean=1) and an extinction (galactic reddening) in the r band of <0.1 magnitudes.

The third section is used to sub-divide the search into three parts based on the psfMag_r value. Three separate runs were carried out at r-band magnitudes:– 14.1 to 16.1, 16.1 to 18.1 and 18.1 to 20.1.

The fourth section constrains the reported magnitudes to values that avoid both excessively bright or faint (and so unreliable) values.

SDSS passband	Brightest	Faintest
u	12.1	21.4
g	14.1	22.6
r	14.1	22.3
i	13.8	21.7
z	12.3	20.1

The final section is used to identify those stars where all five listed de-reddened magnitudes (u, g, r, i and z bands) are within 0.05 magnitudes.

$$\text{AND abs}((s1.psfMag\_u - s1.extinction\_u)-(s2.psfMag\_u - s2.extinction\_u)) < 0.05$$
$$\text{AND abs}((s1.psfMag\_g - s1.extinction\_g)-(s2.psfMag\_g - s2.extinction\_g)) < 0.05$$
$$\text{AND abs}((s1.psfMag\_r - s1.extinction\_r)-(s2.psfMag\_r - s2.extinction\_r)) < 0.05$$
$$\text{AND abs}((s1.psfMag\_i - s1.extinction\_i)-(s2.psfMag\_I - s2.extinction\_i)) < 0.05$$
$$\text{AND abs}((s1.psfMag\_z - s1.extinction\_z)-(s2.psfMag\_z - s2.extinction\_z)) < 0.05$$

Of the 481 "identical twin" systems identified eight already appear in the Washington Visual Double Star Catalog (Mason et al 2001-2014). #57 CRB 81, #71 LDS 6237, #80 GWP 1700, #109 SLW 1053, #147 LDS 3411, #247 BVD 209, #254 LDS 2583 and #262 LDS 5204.

The distribution of the "identical twin" binary star systems is of some interest.

Separation	r-band 14.1 to 16.1	r-band 16.1 to 18.1	r-band 18.1 to 20.1
3 to 10 arc sec	42.3%	24.3%	8.6%
10 to 20 arc sec	27.0%	33.9%	34.4%
20 to 30 arc sec	30.7%	41.8%	57.0%

Fainter stars are, on average, going to be further away from the observer than brighter stars. For any fixed linear separation between the two components the angular separation is going to be reduced as the distance to the star increases.

These results suggest that many of the brighter "identical twin" stars are at separations > 30 arc sec and so outside the analytical range of the data available from SDSS. As progressively fainter – and so more distant stars – are examined more of these pairs are found within the 30 arc sec limit. Similarly bright pairs that are within the 3 to 10 arc sec band when nearby will progressively be found at separations < 3 arc sec as the distance to the star increases.

It is always prudent to check results based on numerical analysis by looking at actual images. There are a number of possible sources including the SDSS DR10 Image List Tool.

http://skyserver.sdss3.org/public/en/tools/chart/listinfo.aspx

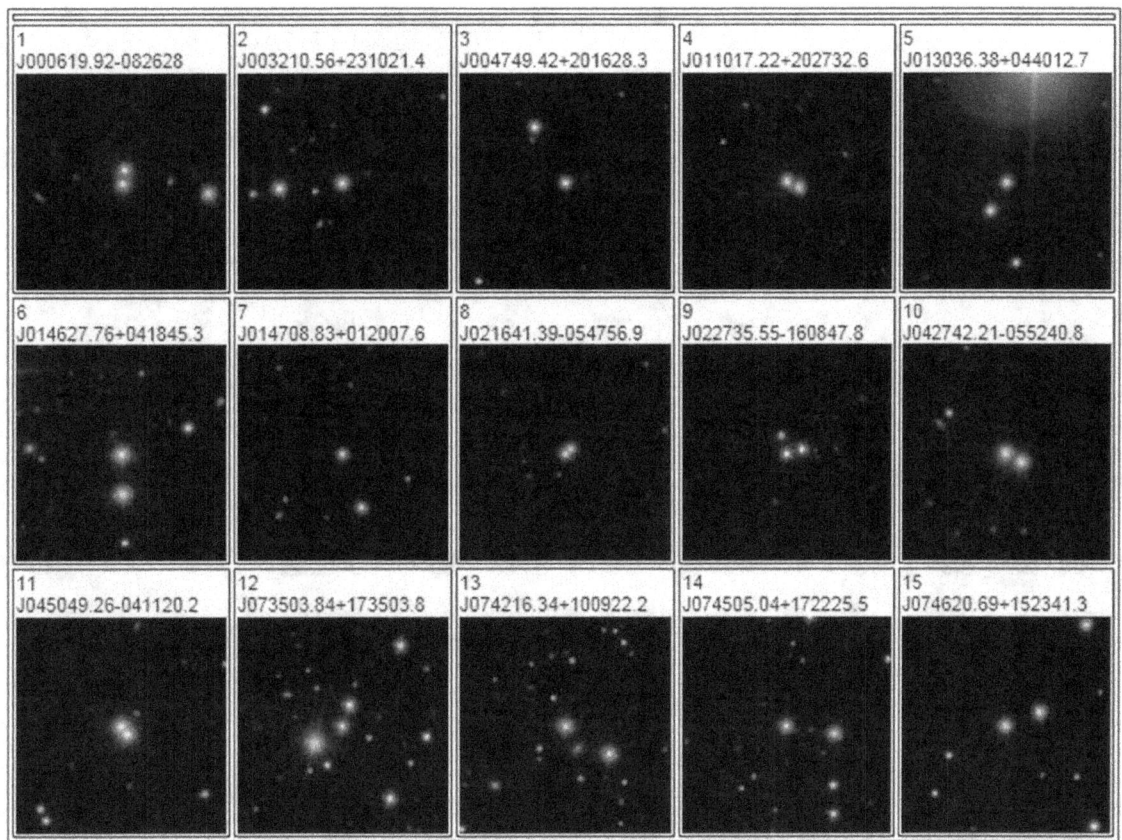

Figure 2 – The first 15 "identical twin" binary star systems

Please note that these are in colour when viewed on the SDSS site.

#	H	M	S	D	M	S	MAG 1	MAG2	SEP	PA
1	0	6	19.92	-8	26	28.03	14.533	14.554	6.071	353.166
2	0	32	10.56	23	10	21.49	14.787	14.826	28.567	94.638
3	0	47	49.43	20	16	28.34	15.723	15.748	28.193	29.212
4	1	10	17.22	20	27	32.64	15.164	15.188	6.024	243.872
5	1	30	36.39	4	40	12.71	15.909	15.916	14.202	150.209
6	1	46	27.77	4	18	45.36	14.023	14.052	17.312	181.458
7	1	47	8.84	1	20	7.69	15.732	15.772	24.749	199.808
8	2	16	41.39	-5	47	56.96	15.788	15.836	3.021	312.456
9	2	27	35.56	-16	8	47.84	15.438	15.465	7.29	284.906
10	4	27	42.22	-5	52	40.86	14.818	14.827	7.972	242.169
11	4	50	49.26	-4	11	20.26	14.421	14.436	4.137	217.685
12	7	35	3.84	17	35	3.82	15.261	15.284	9.845	340.321
13	7	42	16.34	10	9	22.23	14.52	14.538	23.275	238.497
14	7	45	5.04	17	22	25.52	15.035	15.046	22.185	261.46
15	7	46	20.69	15	23	41.39	14.78	14.804	16.681	291.427
16	7	49	24.38	8	1	5.26	14.742	14.745	7.928	297.622
17	7	50	44.46	8	48	18.79	15.323	15.337	5.879	216.505
18	7	55	22.57	7	8	57.61	15.348	15.353	23.463	242.146
19	7	57	59.16	7	7	32.56	15.32	15.343	29.143	128.437
20	7	58	37.23	16	14	22.85	14.697	14.731	4.965	277.081
21	7	59	40.89	0	59	44.77	15.405	15.419	10.848	3.778
22	8	3	57.72	2	51	8.64	15.495	15.516	27.538	55.097
23	8	5	13.42	8	47	50.04	15.8	15.842	11.572	41.88
24	8	5	52.3	7	12	19.99	15.86	15.878	29.333	24.786
25	8	6	33.52	1	15	44.15	15.897	15.91	16.218	317.08
26	8	6	43.29	5	55	24.33	15.825	15.851	29.805	345.171
27	8	7	45.34	-1	12	2.87	15.262	15.264	28.451	287.571
28	8	7	49.85	8	59	46.35	14.986	15.011	21.801	137.376
29	8	8	7.82	12	35	7.45	15.06	15.079	28.558	42.29
30	8	8	22.15	8	2	16.61	15.955	15.957	18.502	177.983
31	8	8	30.5	8	3	6.6	15.293	15.313	26.379	261.635
32	8	8	31.52	2	43	11.62	15.582	15.586	22.516	147.826
33	8	10	31.51	1	29	47.68	15.996	16.003	26.88	141.015
34	8	12	0.97	2	52	24.98	15.58	15.585	9.485	70.013
35	8	17	31.24	8	10	44.09	14.947	14.967	8.721	245.271
36	8	18	54.38	17	20	40.48	15.964	15.971	6.051	221.476
37	8	22	57.3	4	57	42.83	15.505	15.548	27.388	260.1
38	8	24	53.76	4	18	15.25	15.936	15.938	9.804	214.288
39	8	25	42.86	6	47	39.08	14.597	14.62	9.319	227.294
40	8	26	27.35	11	36	32.89	15.629	15.662	3.983	44.68

41	8	26	41.72	9	25	38.81	15.687	15.704	4.983	3.367
42	8	28	22.3	8	7	8.09	14.693	14.733	9.598	214.531
43	8	31	25.89	3	4	45.02	15.491	15.535	10.009	123.332
44	8	32	53.81	-1	2	6.24	14.692	14.715	29.017	180.352
45	8	41	33.92	-2	11	17.77	15.881	15.886	24.132	204.609
46	8	42	53.79	0	46	56.74	15.493	15.504	8.358	258.234
47	8	43	17.29	47	26	8.94	15.417	15.437	3.133	245.028
48	8	43	24.04	4	4	1.24	15.232	15.247	6.533	240.012
49	8	45	49.14	-3	14	9.49	15.315	15.319	9.921	55.013
50	8	47	22.37	12	32	48.26	15.325	15.356	6.928	292.153
51	8	48	43.66	2	2	31.09	15.365	15.371	24.056	105.153
52	8	54	57.93	12	53	40.77	15.796	15.808	6.764	305.142
53	8	55	16.36	19	25	11.96	14.039	14.081	24.237	314.609
54	8	55	59.91	48	51	12.4	15.287	15.318	5.106	167
55	9	2	7.63	23	46	6.93	15.377	15.39	4.11	147.99
56	9	3	43.31	10	36	30.05	15.178	15.207	11.361	11.909
57	9	22	11.39	34	3	26.62	14.894	14.909	13.606	76.799
58	9	26	57.21	30	47	50.2	15.206	15.218	18.655	298.467
59	9	33	12.15	36	3	20.78	15.406	15.435	21.092	99.116
60	9	41	5.82	6	40	28.43	15.342	15.354	21.44	221.923
61	9	41	33.47	10	49	9.94	14.585	14.613	4.875	339.869
62	9	43	41.65	32	29	44.02	15.075	15.099	11.886	273.362
63	9	54	57.68	29	5	23.4	15.611	15.637	24.817	16.48
64	10	17	4.23	26	21	53.7	15.403	15.433	3.934	275.366
65	10	23	14.6	49	20	27.68	16.036	16.076	20.294	291.813
66	10	27	21.82	56	11	45.04	15.015	15.024	11.589	158.548
67	10	31	23.52	15	27	33.11	15.906	15.951	27.196	3.384
68	10	32	22.33	80	11	7.98	15.019	15.033	3.411	127.875
69	10	42	35.31	40	6	29.7	14.455	14.479	3.482	175.706
70	10	56	15.95	52	36	56.9	14.692	14.734	9.689	102.233
71	10	59	1.67	10	55	42.98	15.748	15.759	5.593	253.635
72	11	10	0.03	44	47	20.07	14.463	14.479	3.484	271.532
73	11	11	43.19	21	41	59.13	15.654	15.669	19.402	273.406
74	11	15	11.96	19	56	31.07	15.861	15.889	4.524	178.946
75	11	18	1.92	78	3	23.54	14.89	14.917	5.33	110.349
76	11	27	2.49	25	55	4.6	14.11	14.135	7.06	87.642
77	11	27	36.08	45	15	36.24	15.503	15.54	18.871	11.792
78	11	34	10.07	44	2	35.9	15.457	15.467	13.514	253.752
79	11	38	5.51	65	54	6.44	15.599	15.623	3.876	162.059
80	11	53	28.39	11	46	42.91	14.573	14.609	19.358	138.942
81	11	53	33.73	-2	53	40.3	14.495	14.507	13.26	227.07
82	11	57	52.37	3	17	29.88	14.909	14.936	24.255	16.417
83	12	22	20.53	34	20	48.41	14.775	14.789	9.453	301.143

84	12	28	8.93	61	50	17.86	15.081	15.129	3.549	195.78
85	12	31	16.14	7	21	31.55	15.031	15.034	4.009	52.381
86	12	46	26.93	63	15	55.29	15.741	15.754	9.116	305.938
87	12	49	31.3	-7	12	25.68	15.934	15.944	18.46	160.916
88	12	56	19.56	33	14	12.68	15.919	15.927	25.034	314.269
89	12	59	46.19	53	10	54.11	15.269	15.297	25.753	131.831
90	13	4	57.18	27	7	9.65	14.848	14.871	21.023	291.082
91	13	5	23.94	15	56	10.65	14.141	14.149	5.862	49.839
92	13	10	24.25	33	21	59.99	14.784	14.815	25.373	47.909
93	13	19	54.07	34	44	5.57	15.089	15.091	9.911	121.14
94	13	26	14.18	6	34	35.95	15.359	15.393	23.021	96.153
95	13	45	50.02	2	27	33.09	15.774	15.799	18.074	297.549
96	13	46	46.54	50	52	6.48	15.498	15.498	7.109	114.585
97	13	58	7.8	46	9	13.32	15.249	15.297	23.373	43.385
98	14	3	41.32	29	23	39.64	14.359	14.38	16.547	157.319
99	14	9	6.28	24	40	1.18	15.25	15.28	10.975	94.581
100	14	10	39.21	30	41	33.05	14.698	14.735	4.093	149.808
101	14	12	37.52	16	28	33.69	15.828	15.851	15.872	309.707
102	14	16	14.5	9	42	7.37	16.004	16.025	26.296	201.592
103	14	39	29.49	30	52	49	14.968	15.009	13.012	43.18
104	14	42	36.46	28	6	16.3	15.334	15.352	16.643	337.516
105	14	44	8.87	23	21	2.23	15.289	15.3	18.306	255.176
106	14	48	26.42	9	53	9.25	15.684	15.697	14.58	20.765
107	14	57	8.58	27	17	39.7	15.254	15.277	8.232	281.092
108	15	0	25.26	41	4	3.23	15.396	15.422	9.192	287.148
109	15	2	1.05	60	57	10.97	15.503	15.514	26.828	341.001
110	15	3	29.61	14	31	45.86	15.61	15.653	28.963	27.568
111	15	5	34.87	13	51	11.43	15.708	15.734	4.103	293.984
112	15	7	48.44	47	56	29.53	14.265	14.282	11.104	75.464
113	15	12	14.59	55	37	41.66	16.018	16.031	3.384	94.4
114	15	36	13.1	40	17	14.76	15.997	16.028	4.436	204.087
115	15	40	24.33	29	32	1.54	15.439	15.473	16.955	311.639
116	15	53	55.44	39	7	9.4	15.719	15.755	8.501	173.53
117	16	3	57.42	32	36	36.19	15.502	15.524	15.614	27.596
118	16	20	31.33	31	20	56.71	15.346	15.365	29.234	224.646
119	16	29	43.76	32	59	52.05	15.837	15.841	3.586	312.808
120	16	57	0.27	40	35	46.88	15.943	15.971	19.965	227.774
121	16	57	58.87	35	3	59.18	15.797	15.809	4.885	67.594
122	17	0	40.79	59	49	29.76	15.528	15.544	3.842	262.521
123	17	3	14.12	38	7	54.74	14.197	14.201	12.023	135.334
124	17	6	15.5	65	50	23.83	14.498	14.51	3.081	162.073
125	17	8	23.93	66	19	37.47	15.261	15.262	16.225	88.662
126	17	9	29.87	57	27	56.1	15.473	15.478	10.147	290.351

127	17	11	16.95	60	16	0.05	15.14	15.163	12.035	312.609
128	17	18	36.29	44	26	26.39	15.485	15.499	4.397	255.55
129	17	22	48.83	57	50	26.36	14.423	14.428	5.115	235.757
130	17	31	8.66	63	35	44.03	15.565	15.579	15.528	1.267
131	17	32	21.28	70	55	8.94	15.478	15.492	12.772	246.03
132	17	37	16.52	63	19	28.73	15.309	15.311	26.841	84.022
133	17	38	47.35	62	9	14.62	15.842	15.847	20.93	129.364
134	17	51	59.97	48	32	49.08	15.279	15.299	21.306	139.628
135	17	53	50.17	41	58	29.66	15.268	15.306	22.6	272.231
136	21	46	29.43	-5	47	32.08	15.868	15.91	9.293	60.489
137	23	45	37.85	-10	13	42.79	14.651	14.667	26.328	70.46

Photometry for Batch 1

	Prim u	Prim g	Prim r	Prim i	Priz z	Sec u	Sec g	Sec r	Sec i	Sec z
1	16.680	15.122	14.533	14.312	14.273	16.727	15.125	14.554	14.336	14.296
2	16.604	15.280	14.787	14.616	14.571	16.618	15.320	14.826	14.650	14.598
3	17.406	16.209	15.723	15.604	15.575	17.436	16.220	15.748	15.610	15.575
4	18.889	16.406	15.164	14.667	14.419	18.931	16.436	15.188	14.680	14.434
5	17.317	16.227	15.909	15.800	15.770	17.319	16.229	15.916	15.802	15.763
6	15.772	14.468	14.023	13.957	13.922	15.775	14.502	14.052	13.991	13.958
7	17.129	16.078	15.732	15.627	15.645	17.081	16.098	15.772	15.666	15.686
8	19.880	17.184	15.788	15.145	14.854	19.900	17.216	15.836	15.183	14.888
9	17.606	16.000	15.438	15.260	15.178	17.638	16.030	15.465	15.287	15.206
10	18.755	16.085	14.818	14.328	14.099	18.749	16.095	14.827	14.338	14.104
11	16.016	14.769	14.421	14.313	14.315	16.020	14.768	14.436	14.335	14.343
12	17.074	15.749	15.261	15.122	15.050	17.049	15.760	15.284	15.149	15.081
13	16.032	14.830	14.520	14.443	14.457	16.028	14.871	14.538	14.427	14.427
14	16.344	15.304	15.035	14.955	14.984	16.348	15.317	15.046	14.984	15.007
15	16.099	15.070	14.780	14.713	14.713	16.110	15.081	14.804	14.744	14.737
16	16.103	15.008	14.742	14.639	14.672	16.094	15.027	14.745	14.628	14.646
17	17.436	15.858	15.323	15.102	15.085	17.444	15.873	15.337	15.129	15.103
18	16.964	15.747	15.348	15.271	15.296	16.959	15.746	15.353	15.293	15.324
19	16.738	15.655	15.320	15.216	15.218	16.784	15.667	15.343	15.239	15.254
20	15.922	14.903	14.697	14.660	14.725	15.908	14.933	14.731	14.686	14.749
21	16.824	15.737	15.405	15.320	15.319	16.806	15.769	15.419	15.317	15.307
22	17.220	15.899	15.495	15.365	15.340	17.191	15.920	15.516	15.382	15.343
23	17.696	16.333	15.800	15.628	15.554	17.686	16.371	15.842	15.673	15.591
24	17.212	16.198	15.860	15.754	15.768	17.194	16.184	15.878	15.792	15.815
25	17.684	16.351	15.897	15.767	15.753	17.657	16.357	15.910	15.792	15.778
26	17.245	16.132	15.825	15.762	15.753	17.213	16.147	15.851	15.797	15.780
27	16.716	15.605	15.262	15.189	15.180	16.703	15.606	15.264	15.180	15.169
28	16.492	15.310	14.986	14.910	14.888	16.522	15.347	15.011	14.944	14.925
29	16.801	15.473	15.060	14.930	14.917	16.824	15.504	15.079	14.942	14.914
30	17.518	16.354	15.955	15.844	15.847	17.523	16.368	15.957	15.840	15.832
31	16.735	15.592	15.293	15.227	15.196	16.777	15.621	15.313	15.230	15.228
32	16.897	15.852	15.582	15.499	15.518	16.921	15.861	15.586	15.501	15.509
33	17.494	16.339	15.996	15.857	15.827	17.503	16.330	16.003	15.865	15.851
34	17.001	15.925	15.580	15.507	15.498	17.028	15.923	15.585	15.481	15.454
35	18.899	16.335	14.947	14.258	13.896	18.904	16.357	14.967	14.267	13.901
36	17.424	16.299	15.964	15.832	15.834	17.442	16.309	15.971	15.830	15.828
37	16.980	15.866	15.505	15.394	15.372	16.986	15.883	15.548	15.437	15.415
38	18.376	16.584	15.936	15.759	15.672	18.358	16.543	15.938	15.786	15.711
39	16.272	15.033	14.597	14.482	14.457	16.294	15.060	14.620	14.497	14.475
40	17.314	16.041	15.629	15.487	15.478	17.354	16.081	15.662	15.518	15.492

41	16.943	15.933	15.687	15.650	15.646	16.907	15.922	15.704	15.673	15.679
42	16.215	15.040	14.693	14.619	14.546	16.256	15.078	14.733	14.665	14.593
43	17.584	16.030	15.491	15.306	15.210	17.552	16.061	15.535	15.354	15.255
44	16.055	15.005	14.692	14.626	14.622	16.076	15.001	14.715	14.662	14.653
45	17.159	16.100	15.881	15.759	15.752	17.112	16.114	15.886	15.751	15.730
46	16.988	15.845	15.493	15.372	15.386	17.013	15.854	15.504	15.380	15.390
47	19.459	16.810	15.417	14.296	13.714	19.507	16.830	15.437	14.313	13.693
48	17.550	15.821	15.232	15.084	15.027	17.594	15.844	15.247	15.094	15.046
49	16.685	15.670	15.315	15.214	15.206	16.664	15.630	15.319	15.216	15.231
50	17.339	15.836	15.325	15.146	15.124	17.351	15.875	15.356	15.180	15.148
51	16.926	15.740	15.365	15.235	15.261	16.922	15.740	15.371	15.250	15.272
52	17.618	16.295	15.796	15.609	15.574	17.626	16.306	15.808	15.621	15.577
53	16.054	14.560	14.039	13.893	13.881	16.066	14.595	14.081	13.927	13.917
54	19.024	16.528	15.287	14.677	14.371	19.052	16.564	15.318	14.697	14.380
55	17.683	15.997	15.377	15.159	15.101	17.689	16.014	15.390	15.175	15.117
56	16.776	15.586	15.178	15.057	15.007	16.811	15.612	15.207	15.084	15.033
57	18.770	16.238	14.894	13.935	13.414	18.794	16.258	14.909	13.949	13.435
58	16.692	15.560	15.206	15.069	15.056	16.730	15.573	15.218	15.064	15.061
59	17.276	15.927	15.406	15.255	15.246	17.294	15.942	15.435	15.296	15.286
60	17.439	15.923	15.342	15.127	15.052	17.405	15.898	15.354	15.157	15.088
61	16.036	14.879	14.585	14.447	14.463	16.036	14.902	14.613	14.485	14.510
62	16.662	15.488	15.075	14.972	14.941	16.661	15.482	15.099	15.004	14.961
63	17.495	16.071	15.611	15.465	15.453	17.521	16.112	15.637	15.465	15.450
64	19.410	16.759	15.403	14.322	13.741	19.403	16.790	15.433	14.352	13.762
65	18.238	16.691	16.036	15.830	15.725	18.271	16.722	16.076	15.864	15.765
66	16.916	15.546	15.015	14.818	14.773	16.920	15.546	15.024	14.841	14.812
67	17.658	16.383	15.906	15.766	15.700	17.693	16.430	15.951	15.795	15.719
68	16.564	15.359	15.019	14.900	14.910	16.585	15.383	15.033	14.910	14.923
69	16.839	15.064	14.455	14.253	14.135	16.845	15.082	14.479	14.279	14.155
70	16.098	15.036	14.692	14.584	14.591	16.135	15.069	14.734	14.612	14.612
71	19.907	17.126	15.748	14.357	13.570	19.911	17.144	15.759	14.361	13.568
72	15.960	14.866	14.463	14.367	14.360	15.987	14.881	14.479	14.384	14.374
73	17.726	16.218	15.654	15.494	15.430	17.719	16.250	15.669	15.519	15.451
74	19.743	17.238	15.861	14.555	13.858	19.706	17.242	15.889	14.569	13.907
75	16.489	15.255	14.890	14.792	14.804	16.484	15.275	14.917	14.813	14.808
76	15.970	14.549	14.110	13.921	13.920	15.970	14.573	14.135	13.944	13.935
77	17.021	15.892	15.503	15.394	15.363	16.985	15.901	15.540	15.435	15.407
78	16.992	15.827	15.457	15.336	15.303	16.958	15.871	15.467	15.337	15.299
79	18.000	16.254	15.599	15.390	15.289	18.040	16.278	15.623	15.413	15.316
80	15.730	14.790	14.573	14.477	14.490	15.761	14.828	14.609	14.518	14.537
81	15.857	14.864	14.495	14.396	14.353	15.877	14.889	14.507	14.412	14.369
82	16.394	15.291	14.909	14.782	14.772	16.428	15.321	14.936	14.814	14.809
83	17.266	15.396	14.775	14.614	14.533	17.267	15.420	14.789	14.627	14.561

84	19.083	16.493	15.081	13.988	13.363	19.121	16.540	15.129	14.026	13.389
85	16.786	15.471	15.031	14.899	14.885	16.819	15.478	15.034	14.903	14.898
86	17.307	16.153	15.741	15.638	15.619	17.322	16.166	15.754	15.655	15.642
87	18.035	16.512	15.934	15.723	15.658	17.985	16.536	15.944	15.708	15.616
88	17.423	16.317	15.919	15.773	15.733	17.445	16.330	15.927	15.781	15.752
89	18.022	16.065	15.269	15.006	14.871	18.032	16.085	15.297	15.024	14.881
90	16.659	15.333	14.848	14.678	14.639	16.653	15.328	14.871	14.714	14.683
91	16.049	14.597	14.141	13.978	13.901	16.037	14.602	14.149	13.994	13.933
92	16.416	15.196	14.784	14.679	14.640	16.399	15.199	14.815	14.720	14.674
93	16.381	15.405	15.089	14.980	14.951	16.371	15.397	15.091	14.989	14.970
94	16.806	15.733	15.359	15.220	15.209	16.770	15.772	15.393	15.232	15.181
95	17.259	16.173	15.774	15.637	15.632	17.300	16.196	15.799	15.678	15.679
96	18.582	16.303	15.498	15.211	15.048	18.550	16.298	15.498	15.204	15.048
97	19.315	16.647	15.249	14.340	13.892	19.275	16.666	15.297	14.362	13.932
98	16.943	14.982	14.359	14.218	14.147	16.972	14.996	14.380	14.237	14.166
99	17.977	15.998	15.250	14.960	14.853	17.973	16.029	15.280	14.971	14.831
100	16.766	15.201	14.698	14.530	14.449	16.812	15.242	14.735	14.569	14.484
101	17.369	16.192	15.828	15.707	15.662	17.376	16.198	15.851	15.749	15.705
102	17.388	16.353	16.004	15.878	15.878	17.372	16.365	16.025	15.921	15.908
103	18.543	15.974	14.968	14.635	14.446	18.532	16.024	15.009	14.667	14.475
104	17.585	15.973	15.334	15.135	15.036	17.546	15.976	15.352	15.176	15.072
105	17.004	15.749	15.289	15.159	15.120	16.985	15.737	15.300	15.174	15.157
106	17.390	16.144	15.684	15.497	15.505	17.437	16.141	15.697	15.523	15.505
107	16.869	15.658	15.254	15.121	15.104	16.874	15.677	15.277	15.132	15.132
108	18.957	16.515	15.396	14.971	14.788	19.000	16.540	15.422	14.994	14.790
109	19.529	16.910	15.503	14.515	13.977	19.537	16.914	15.514	14.520	13.984
110	17.220	16.004	15.610	15.488	15.460	17.225	16.046	15.653	15.528	15.493
111	19.366	16.839	15.708	15.302	15.108	19.381	16.888	15.734	15.321	15.119
112	17.747	15.285	14.265	13.928	13.767	17.791	15.311	14.282	13.948	13.773
113	19.803	17.125	16.018	15.581	15.387	19.758	17.154	16.031	15.585	15.390
114	18.124	16.549	15.997	15.798	15.703	18.107	16.570	16.028	15.846	15.746
115	17.043	15.859	15.439	15.323	15.285	17.017	15.878	15.473	15.357	15.329
116	17.524	16.115	15.719	15.561	15.522	17.513	16.139	15.755	15.582	15.517
117	18.168	16.242	15.502	15.187	15.065	18.190	16.267	15.524	15.198	15.061
118	18.306	16.181	15.346	15.089	14.956	18.259	16.200	15.365	15.080	14.934
119	17.482	16.216	15.837	15.681	15.654	17.436	16.201	15.841	15.684	15.656
120	17.388	16.272	15.943	15.846	15.811	17.428	16.320	15.971	15.865	15.837
121	17.714	16.286	15.797	15.632	15.599	17.735	16.305	15.809	15.640	15.612
122	17.029	15.929	15.528	15.393	15.347	17.073	15.956	15.544	15.402	15.358
123	15.627	14.577	14.197	14.108	14.144	15.640	14.576	14.201	14.104	14.134
124	16.009	14.890	14.498	14.398	14.413	16.029	14.914	14.510	14.398	14.383
125	17.112	15.780	15.261	15.100	15.066	17.074	15.785	15.262	15.117	15.049
126	17.248	15.924	15.473	15.338	15.293	17.205	15.899	15.478	15.378	15.313

127	17.244	15.680	15.140	15.001	14.938	17.197	15.701	15.163	15.021	14.971
128	17.138	15.891	15.485	15.364	15.307	17.179	15.903	15.499	15.389	15.349
129	16.381	14.925	14.423	14.258	14.193	16.336	14.917	14.428	14.272	14.202
130	19.444	16.872	15.565	14.429	13.834	19.484	16.882	15.579	14.432	13.836
131	17.971	16.102	15.478	15.286	15.216	17.923	16.087	15.492	15.314	15.250
132	16.992	15.714	15.309	15.162	15.137	17.016	15.756	15.311	15.141	15.102
133	17.763	16.324	15.842	15.706	15.621	17.795	16.316	15.847	15.718	15.625
134	16.862	15.668	15.279	15.131	15.115	16.898	15.692	15.299	15.134	15.107
135	17.435	15.811	15.268	15.090	15.022	17.397	15.842	15.306	15.119	15.040
136	17.360	16.268	15.868	15.773	15.735	17.365	16.306	15.910	15.793	15.746
137	17.164	15.337	14.651	14.444	14.339	17.154	15.341	14.667	14.440	14.321

Batch 2 – where s1.psfMag_r between 16.1 and 18.1

#	H	M	S	D	M	S	MAG 1	MAG2	SEP	PA
138	0	14	56.91	27	11	26.6	16.417	16.43	29.98	91.515
139	0	40	38.33	25	11	30.82	16.136	16.152	24.414	72.492
140	0	44	26.45	3	8	51.41	16.493	16.503	23.694	29.571
141	0	48	48.89	1	15	44.2	16.71	16.744	8.04	328.255
142	1	9	19.22	0	54	19.08	16.352	16.366	29.734	77.621
143	1	12	4.89	3	16	48.97	16.237	16.24	4.608	82.672
144	1	39	35.63	-1	59	5.72	16.344	16.362	17.193	105.698
145	1	43	27.73	33	58	17.41	16.832	16.833	27.794	47.912
146	2	22	5.64	3	25	29.17	16.416	16.434	11.571	148.985
147	2	55	52.36	-15	20	47.25	16.332	16.365	18.296	23.253
148	7	37	20.69	11	5	15.5	17.139	17.15	6.366	45.644
149	7	44	39.65	8	31	49.35	17.879	17.917	24.506	238.323
150	7	44	59.6	12	20	2.45	16.065	16.082	16.521	65.843
151	7	46	46.3	11	27	25.72	17.777	17.795	25.057	325.851
152	7	47	56.64	13	0	22.07	17.815	17.833	11.408	244.738
153	7	48	10.93	15	38	22.03	16.853	16.874	28.678	29.546
154	7	48	28.26	8	52	59.08	16.163	16.172	24.838	199.539
155	7	48	30.13	13	3	34.47	16.851	16.89	4.178	349.279
156	7	48	44.25	28	45	8.83	17.574	17.608	28.362	232.601
157	7	49	23.06	6	47	16.5	17.532	17.57	26.334	190.058
158	7	50	5.91	6	43	57.86	16.292	16.315	3.407	222.447
159	7	50	30.52	6	55	40	16.655	16.662	17.88	349.826
160	7	50	32.64	7	7	16.9	17.509	17.536	29.599	34.66
161	7	50	37.97	13	6	40.6	16.982	17.005	19.836	249.156
162	7	50	52.85	13	42	51.52	16.642	16.682	23.675	74.951
163	7	51	10.33	13	32	12.75	16.293	16.296	4.195	175.981
164	7	51	32.53	12	55	1.55	17.465	17.507	24.909	34.453
165	7	52	40.74	6	50	32.26	17.377	17.396	23.127	240.722
166	7	52	45.24	8	42	49.39	16.803	16.823	21.197	331.946
167	7	53	0.91	26	59	3.1	16.36	16.409	23.057	299.284
168	7	53	16.56	7	43	26.27	16.151	16.158	23.658	282.907
169	7	53	31.93	11	2	25.78	17.466	17.475	10.047	29.644
170	7	54	10.78	5	57	29.43	16.911	16.95	4.846	54.796
171	7	54	17.84	-1	48	55.12	16.86	16.87	19.967	302.955
172	7	54	37.2	4	56	55.28	17.783	17.803	17.087	93.487
173	7	54	56.68	5	39	1.77	17.016	17.038	25.383	278.138
174	7	55	0.33	5	23	33.12	16.692	16.7	28.465	203.437
175	7	57	19.76	13	37	29.42	17.636	17.676	17.355	33.902
176	7	58	36.1	6	15	57.13	17.536	17.545	17.526	328.154
177	7	59	10.38	0	2	39.75	17.073	17.108	27.817	327.362

178	8	0	36.83	11	48	41.97	16.958	16.992	24.201	231.841
179	8	1	11.49	4	18	3.15	17.31	17.313	20.351	288.411
180	8	1	20.25	10	19	9.52	16.352	16.382	22.061	190.066
181	8	2	7.38	2	11	16.57	16.396	16.41	24.969	16.744
182	8	2	31.44	2	39	6.84	17.527	17.535	29.908	4.161
183	8	3	8.41	-2	12	24.15	17.369	17.4	8.253	225.569
184	8	3	24.99	3	49	16.75	17.459	17.468	21.709	13.741
185	8	4	1.59	-1	39	18.63	17.496	17.533	20.446	204.348
186	8	4	6.66	-1	20	2.57	16.54	16.584	16.355	239.491
187	8	4	22.29	7	14	3.99	16.173	16.195	15.936	304.599
188	8	4	39.91	16	7	20.54	17.757	17.778	23.754	76.645
189	8	5	6.68	-1	37	56.19	17.25	17.278	23.051	325.002
190	8	5	37.79	11	42	10.17	17.107	17.118	19.141	113.906
191	8	5	47.42	-1	56	21.33	17.043	17.045	19.513	11.978
192	8	6	21.1	-2	18	19.09	17.751	17.77	21.51	231.446
193	8	6	42.17	25	57	6.87	16.637	16.662	18.717	328.725
194	8	6	58.04	1	2	20.78	16.872	16.889	27.47	96.211
195	8	7	12.38	-1	52	55.36	17.567	17.578	23.691	163.618
196	8	7	15.03	0	57	2.06	17.943	17.949	25.092	51.556
197	8	7	17.98	15	32	17.54	17.3	17.312	3.164	61.97
198	8	8	16.96	8	43	38.27	17.338	17.367	27.167	11.282
199	8	9	34.41	5	38	15.02	16.025	16.059	29.4	148.232
200	8	11	28.4	11	55	19.77	16.968	16.991	14.96	300.505
201	8	12	24.48	8	15	31.93	16.298	16.312	20.239	87.794
202	8	16	29.28	21	1	21.78	17.67	17.694	16.354	52.323
203	8	18	32.21	3	17	52.83	16.236	16.241	22.878	27.66
204	8	19	4.94	17	30	45.96	16.525	16.526	17.235	263.746
205	8	19	52.92	11	41	35.31	16.082	16.084	7.745	331.659
206	8	20	43.62	13	6	44.41	17.031	17.058	4.668	260.181
207	8	22	35.28	4	17	56.24	17.075	17.091	23.3	332.456
208	8	24	31.58	25	20	59.74	17.158	17.176	17.357	289.062
209	8	25	44.41	-4	13	44.22	16.171	16.182	20.35	190.201
210	8	30	19.01	-3	17	32.75	16.967	16.994	22.18	79.239
211	8	33	59.53	42	11	51.86	17.179	17.223	3.24	152.04
212	8	34	15.19	23	37	50.01	16.523	16.532	5.689	166.376
213	8	34	42.84	12	23	5.72	17.895	17.907	17.661	326.106
214	8	35	31.82	45	26	3.39	16.06	16.073	26.559	223.306
215	8	36	1.59	20	32	35.2	17.227	17.246	11.505	130.532
216	8	38	1.91	-4	48	58.42	16.23	16.254	13.361	353.863
217	8	38	10.61	16	27	9.77	16.905	16.923	17.06	120.287
218	8	38	30.26	-2	15	3.65	16.855	16.862	25.63	19.032
219	8	39	29.6	38	38	34.59	16.552	16.577	3.441	24.654
220	8	40	33.84	3	53	2.46	17.558	17.571	15.043	23.245

221	8	41	6.94	6	32	12.39	17.336	17.339	25.255	62.016
222	8	43	29.79	-1	22	48.24	16.728	16.744	16.126	335.988
223	8	47	57.9	-2	17	23.69	16.608	16.631	7.206	45.395
224	8	48	35.08	19	15	49.21	17.41	17.418	19.968	172.62
225	8	50	58.07	82	37	25.53	16.849	16.854	23.159	233.181
226	8	57	1.52	24	12	52.19	17.531	17.566	21.599	82.262
227	8	58	24.1	16	50	22.33	17.594	17.602	25.49	47.874
228	9	2	5.09	30	16	8	16.475	16.486	5.841	25.183
229	9	2	22.77	11	58	40.18	16.674	16.707	11.982	287.482
230	9	4	10.98	40	42	17.58	17.293	17.314	10.394	60.426
231	9	11	10.97	15	56	36.32	17.794	17.798	19.825	77.221
232	9	14	39.61	31	55	27.04	17.474	17.493	4.573	318.314
233	9	18	29.72	42	37	35.07	16.892	16.91	21.535	124.597
234	9	21	12.79	17	10	43.76	16.448	16.477	11.371	174.642
235	9	25	6.96	-1	34	5.09	16.426	16.471	7.254	177.609
236	9	26	8.14	30	53	28.49	16.955	16.995	17.507	344.15
237	9	29	51.48	30	35	14.13	17.092	17.121	23.948	38.649
238	9	32	29.97	22	41	33.16	16.598	16.623	17.865	318.809
239	9	34	12.12	12	49	8.48	16.641	16.656	29.44	269.325
240	9	35	30.14	22	49	43.63	16.211	16.229	10.184	7.438
241	9	38	4.6	15	38	55.76	16.036	16.063	5.176	256.584
242	9	41	17.58	60	49	15.38	17.963	17.971	3.067	286.45
243	9	42	22.86	32	0	11.68	16.457	16.468	22.521	54.62
244	9	47	54.7	28	27	21.99	17.169	17.169	3.735	233.06
245	9	52	15.91	43	17	27.46	16.251	16.27	12.548	115.148
246	9	53	15.43	81	29	0.59	16.181	16.183	28.884	168.571
247	9	56	11.47	46	16	49.38	16.381	16.412	8.613	327.266
248	10	1	40.34	2	8	5.31	17.362	17.367	29.595	244.164
249	10	8	35.85	37	47	29.52	17.133	17.159	4.356	357.762
250	10	12	21.44	23	11	50.58	16.158	16.159	17.111	35.581
251	10	15	51.23	62	3	2.53	16.827	16.832	9.818	29.716
252	10	17	36.9	-3	15	15.93	16.285	16.297	6.485	38.139
253	10	23	33.46	22	54	17.08	16.332	16.345	3.037	298.601
254	10	26	36.91	62	45	18.51	17.351	17.398	9.739	312.792
255	10	28	40.96	59	55	29.61	16.153	16.182	13.641	175.684
256	10	29	49.73	9	15	21.58	17.31	17.315	28.476	250.109
257	10	32	33.12	9	58	1.39	16.934	16.943	3.486	71.476
258	10	42	31.01	4	59	55.17	17.646	17.656	10.133	348.591
259	10	42	57.91	28	56	4.16	17.634	17.655	26.274	161.583
260	11	7	43.31	7	56	24.99	16.088	16.116	3.822	356.48
261	11	15	43.27	25	37	1.07	17.1	17.116	6.421	234.409
262	11	21	33.6	51	12	0.07	16.227	16.242	4.029	272.431
263	11	26	53	68	0	52.18	16.097	16.111	3.001	62.858

264	11	31	31.32	39	56	6.64	17.981	18.004	24.26	164.615
265	11	33	44.11	-3	13	10.42	16.174	16.187	8.998	133.252
266	11	38	38.35	68	22	52.18	16.927	16.932	4.556	349.021
267	11	44	55.63	4	39	4.28	17.723	17.724	5.996	24.511
268	11	53	36.27	29	9	38.48	17.301	17.332	14.679	341.687
269	12	4	16.38	-1	12	13.83	17.151	17.169	19.352	221.818
270	12	12	4.79	58	31	45.4	16.588	16.593	13.915	32.774
271	12	16	4.57	0	13	42.21	17.53	17.565	22.425	129.151
272	12	24	47.11	-2	26	38.34	16.289	16.305	12.733	278.994
273	12	39	13.04	11	10	18.55	17.732	17.757	3.867	250.201
274	12	40	24.42	59	7	5.85	16.203	16.219	4.054	302.179
275	12	40	44.84	-10	7	56.28	17.08	17.102	12.831	314.65
276	12	45	6.73	13	11	46.21	16.066	16.085	5.017	299.469
277	12	49	59.96	8	30	56.76	17.717	17.758	15.765	335.946
278	12	52	17.84	7	16	55.98	17.62	17.66	14.68	313.497
279	12	55	21.3	63	31	6.85	16.316	16.332	8.162	321.806
280	13	2	13.07	0	12	3.31	16.244	16.28	29.958	155.589
281	13	6	42.52	19	32	31.08	16.501	16.502	9.169	329.052
282	13	9	55.9	33	12	10.21	17.492	17.53	11.32	206.195
283	13	10	31.38	10	33	53.41	17.429	17.453	3.023	170.554
284	13	19	25.87	45	1	27.77	16.713	16.738	5.933	339.17
285	13	20	37.51	53	30	49.4	16.207	16.21	28.782	136.774
286	13	24	2.29	59	15	59.1	18.051	18.08	10.29	278.948
287	13	27	4.69	32	24	27.44	16.554	16.577	3.91	211.17
288	13	34	6.13	23	3	27.36	17.249	17.28	27.146	53.073
289	13	37	34.53	26	33	27.42	16.382	16.407	12.608	355.78
290	13	39	19.24	19	32	8.64	16.487	16.501	13.149	290.919
291	13	44	29.08	6	47	51.43	16.611	16.627	23.217	224.913
292	13	48	36.6	13	22	28.29	17.935	17.959	25.428	103.232
293	13	51	43.99	8	26	55.02	17.615	17.65	25.755	60.675
294	14	2	5.49	36	37	11.88	17.435	17.474	5.523	121.646
295	14	7	29.7	3	53	37.41	17.089	17.129	18.665	103.339
296	14	11	25.07	5	10	26.72	17.773	17.794	23.877	43.146
297	14	14	15.72	17	8	14.76	16.737	16.744	14.788	83.34
298	14	18	23.41	7	39	54.7	16.134	16.148	5.786	238.211
299	14	23	50.59	17	26	6.59	17.416	17.428	22.398	41.235
300	14	27	43.58	32	12	16.36	17.932	17.939	22.774	55.951
301	14	31	15.99	4	10	11.3	16.739	16.785	19.268	254.674
302	14	31	40.95	2	24	7.58	16.914	16.94	26.496	208.732
303	14	35	0.31	61	33	12.28	17.943	17.946	22.548	146.971
304	14	39	11.5	15	45	38.29	16.3	16.327	16.536	352.133
305	14	40	16.87	47	2	3.98	17.825	17.839	20.565	61.589
306	14	43	46.14	8	19	58.55	17.174	17.194	18.903	327.487

307	14	45	25.89	9	45	53.32	17.154	17.156	27.628	34.668
308	14	46	3.12	41	17	39.39	16.42	16.422	6.035	146.021
309	14	49	15.64	48	6	37.63	16.509	16.517	29.774	347.432
310	14	50	3.61	8	36	30.6	17.569	17.593	18.494	34.465
311	14	50	17.52	33	46	0.86	16.423	16.449	29.815	123.513
312	14	54	11.36	16	43	34.83	16.179	16.192	13.953	271.617
313	14	56	56.43	8	10	35.78	16.875	16.915	4.346	104.343
314	15	8	45.78	32	41	16.55	16.551	16.564	19.146	331.008
315	15	17	5.93	9	11	17.31	16.98	16.998	27.631	195.152
316	15	19	26.71	52	25	9.99	16.461	16.494	9.721	288.509
317	15	20	30.99	44	35	29.76	17.653	17.701	19.524	305.583
318	15	21	36.11	28	48	40.76	17.508	17.541	23.378	128.805
319	15	22	16.04	32	5	43.02	16.104	16.124	4.227	44.268
320	15	22	23.62	15	46	8.45	16.808	16.827	25.706	152.175
321	15	29	52.6	42	12	30.57	16.983	16.991	29.834	198.7
322	15	31	50.57	32	32	14.9	16.361	16.378	13.114	0.477
323	15	42	36.83	45	20	10.56	16.911	16.917	21.842	240.347
324	15	53	38.01	34	21	13.79	16.585	16.596	26.518	354.889
325	15	53	45.98	42	58	14.11	16.743	16.754	3.38	243.414
326	15	53	52.42	49	40	37.21	16.955	16.962	19.258	18.48
327	15	57	7.86	56	57	54.84	17.003	17.006	7.667	42.154
328	16	7	56.76	43	14	16.19	17.337	17.367	23.538	297.049
329	16	8	44.21	35	24	55.04	17.448	17.448	24.135	284.423
330	16	13	23.6	42	52	36.59	16.521	16.537	17.905	18.024
331	16	14	32.54	37	17	47.11	17.601	17.628	24.765	78.315
332	16	14	43.06	56	53	53.2	17.667	17.667	18.468	234.705
333	16	19	34.52	36	27	10.24	16.89	16.898	24.417	265.783
334	16	21	18.72	46	15	3.78	16.549	16.553	24.922	276.409
335	16	22	34.87	29	15	49.13	17.819	17.83	19.983	136.228
336	16	29	0.1	57	7	44.39	16.233	16.243	11.332	177.833
337	16	29	34.9	45	32	25.33	18.002	18.025	17.223	278.119
338	16	30	32.09	31	26	16.21	16.45	16.48	19.166	11.688
339	16	31	14.2	30	17	6.96	16.813	16.828	13.485	6.733
340	16	33	35.02	63	32	21.85	16.811	16.836	13.395	58.296
341	16	34	5.78	29	24	58.39	17.102	17.115	27.005	40.647
342	16	34	22.1	38	40	43.4	17.486	17.519	8.857	150.684
343	16	34	46.17	43	44	35.16	17.435	17.442	21.836	62.443
344	16	36	25.32	38	48	2.7	17.545	17.568	3.868	94.166
345	16	37	49.53	33	14	15.14	17.585	17.586	12.28	46.279
346	16	38	1.21	29	49	1.7	17.831	17.854	17.095	201.419
347	16	38	4.29	37	50	36.75	18.003	18.02	29.461	100.049
348	16	45	16.82	39	39	10.49	16.652	16.686	25.877	57.676
349	16	50	53.98	31	16	3.19	16.896	16.905	20.013	307.214

350	16	57	17.32	35	5	6.32	16.666	16.695	17.142	232.864
351	16	58	13.61	46	56	24.39	17.914	17.918	5.596	235.403
352	17	0	43.47	61	41	3.1	16.969	16.979	8.006	58.09
353	17	1	13.68	40	6	34.88	17.719	17.741	15.102	146.278
354	17	1	15.86	42	10	0.12	16.069	16.088	16.802	50.272
355	17	2	52.66	49	11	49.95	17.054	17.092	16.554	37.001
356	17	3	6.42	58	56	11.53	17.585	17.613	13.939	208.702
357	17	3	18.38	33	25	32.95	17.156	17.178	20.805	141.393
358	17	5	39.06	76	37	54.63	17.99	18.006	26.035	239.38
359	17	7	38.5	61	4	34.71	17.287	17.301	5.222	64.093
360	17	13	17.38	34	12	21.34	17.634	17.667	20.49	89.036
361	17	14	29.21	35	1	34.05	16.203	16.221	21.913	245.766
362	17	17	15.75	68	49	44.2	17.27	17.288	29.535	286.306
363	17	18	1.21	35	10	41.51	17.378	17.406	4.276	312.363
364	17	18	51.63	43	24	28.9	16.923	16.936	14.762	274.897
365	17	19	10.3	42	42	18.89	16.983	16.986	27.841	100.438
366	17	19	26.13	34	47	14.68	17.79	17.798	22.28	25.562
367	17	20	53.47	52	21	22.32	17.567	17.57	13.328	209.581
368	17	26	52.71	43	3	47.22	17.45	17.485	17.136	330.218
369	17	28	32.29	32	59	2.94	16.079	16.086	19.09	275.352
370	17	29	30.93	71	5	5.12	16.415	16.459	12.188	292.062
371	17	32	28.12	44	47	56.59	18.04	18.057	27.843	302.066
372	17	34	54.83	35	14	41.25	16.893	16.908	8.877	284.34
373	17	35	16.22	52	3	2.72	16.473	16.521	28.629	109.274
374	17	38	2.15	62	45	6.24	16.936	16.939	22.236	108.804
375	17	41	26.27	52	41	52.83	17.139	17.141	18.469	91.415
376	17	44	5.75	47	11	22.61	17.427	17.439	23.09	62.663
377	17	47	39.81	47	16	41.11	17.766	17.775	19.917	268.044
378	17	52	47.51	38	22	56.7	16.362	16.378	22.39	122.196
379	17	54	24.44	60	18	26.62	17.775	17.805	19.487	199.37
380	17	59	44.48	42	5	56.82	16.99	16.991	3.024	129.541
381	18	5	29.58	42	31	5.1	16.645	16.649	29.55	215.924
382	18	17	3.05	41	19	19.56	17.599	17.629	12.138	9.843
383	20	34	22.84	-6	31	48.36	17.694	17.703	8.064	259.508
384	21	32	11.03	-8	37	1.15	16.799	16.805	23.623	143.302
385	21	49	12.09	-3	43	11.11	16.582	16.591	24.788	105.817
386	23	32	58.69	-7	17	43.45	16.896	16.904	26.527	355.239
387	23	40	24.25	-2	30	46.58	16.798	16.825	8.407	228.674
388	23	54	35.11	-6	5	53.2	17.091	17.096	29.184	129.883

Photometry for Batch 2

	Prim u	Prim g	Prim r	Prim i	Prim z	Sec u	Sec g	Sec r	Sec i	Sec z
138	18.055	16.821	16.417	16.316	16.241	18.093	16.868	16.430	16.301	16.223
139	17.728	16.600	16.136	15.980	15.961	17.745	16.585	16.152	16.003	15.983
140	18.486	17.021	16.493	16.335	16.245	18.456	17.018	16.503	16.350	16.244
141	20.718	18.099	16.710	15.975	15.622	20.751	18.118	16.744	15.999	15.651
142	17.730	16.706	16.352	16.240	16.208	17.740	16.734	16.366	16.238	16.186
143	17.539	16.567	16.237	16.129	16.101	17.556	16.586	16.240	16.128	16.101
144	20.162	17.689	16.344	15.786	15.449	20.172	17.683	16.362	15.804	15.465
145	18.166	17.143	16.832	16.708	16.710	18.139	17.111	16.833	16.727	16.739
146	20.431	17.916	16.416	15.136	14.500	20.406	17.942	16.434	15.136	14.491
147	20.210	17.747	16.332	15.700	15.338	20.176	17.783	16.365	15.738	15.361
148	18.584	17.508	17.139	17.017	16.977	18.632	17.536	17.150	17.052	17.014
149	20.080	18.439	17.879	17.774	17.534	20.095	18.480	17.917	17.823	17.571
150	17.417	16.351	16.065	16.002	16.007	17.452	16.363	16.082	16.042	16.036
151	19.662	18.248	17.777	17.583	17.552	19.651	18.263	17.795	17.609	17.567
152	19.758	18.346	17.815	17.722	17.627	19.762	18.369	17.833	17.740	17.642
153	18.299	17.186	16.853	16.774	16.762	18.265	17.196	16.874	16.798	16.756
154	19.093	16.925	16.163	15.916	15.837	19.083	16.924	16.172	15.930	15.841
155	18.665	17.375	16.851	16.712	16.686	18.712	17.376	16.890	16.720	16.730
156	19.621	18.096	17.574	17.366	17.243	19.643	18.130	17.608	17.364	17.273
157	20.269	18.259	17.532	17.295	17.189	20.253	18.295	17.570	17.303	17.187
158	18.057	16.739	16.292	16.127	16.112	18.086	16.777	16.315	16.127	16.087
159	18.724	17.264	16.655	16.487	16.420	18.762	17.240	16.662	16.490	16.419
160	21.067	18.634	17.509	17.144	16.945	21.050	18.653	17.536	17.140	16.924
161	18.704	17.426	16.982	16.858	16.816	18.722	17.454	17.005	16.865	16.808
162	18.201	17.019	16.642	16.484	16.450	18.199	17.049	16.682	16.529	16.475
163	17.921	16.695	16.293	16.192	16.084	17.884	16.679	16.296	16.207	16.115
164	19.057	17.850	17.465	17.328	17.275	19.017	17.881	17.507	17.359	17.324
165	18.964	17.756	17.377	17.206	17.269	18.982	17.771	17.396	17.245	17.241
166	18.443	17.226	16.803	16.650	16.605	18.467	17.224	16.823	16.642	16.608
167	17.808	16.714	16.360	16.222	16.232	17.768	16.733	16.409	16.250	16.267
168	17.781	16.565	16.151	15.982	15.945	17.808	16.560	16.158	15.986	15.925
169	19.609	18.013	17.466	17.229	17.173	19.576	18.026	17.475	17.236	17.173
170	18.441	17.296	16.911	16.807	16.793	18.432	17.324	16.950	16.855	16.793
171	18.232	17.202	16.860	16.756	16.776	18.203	17.166	16.870	16.775	16.811
172	19.252	18.166	17.783	17.660	17.666	19.226	18.151	17.803	17.699	17.639
173	18.465	17.383	17.016	16.905	16.897	18.452	17.364	17.038	16.932	16.928
174	18.088	17.024	16.692	16.592	16.587	18.051	16.996	16.700	16.619	16.602
175	19.038	17.996	17.636	17.541	17.540	19.062	18.010	17.676	17.582	17.585
176	19.164	17.934	17.536	17.403	17.384	19.122	17.959	17.545	17.404	17.359
177	18.533	17.403	17.073	16.889	16.877	18.527	17.423	17.108	16.933	16.903

178	19.252	17.582	16.958	16.721	16.614	19.204	17.584	16.992	16.708	16.649
179	19.731	17.960	17.310	17.106	17.021	19.707	17.967	17.313	17.102	16.983
180	17.787	16.696	16.352	16.232	16.182	17.819	16.727	16.382	16.244	16.184
181	17.901	16.745	16.396	16.265	16.214	17.857	16.794	16.410	16.298	16.216
182	20.067	18.240	17.527	17.282	17.159	20.025	18.279	17.535	17.272	17.116
183	18.951	17.753	17.369	17.250	17.210	19.000	17.777	17.400	17.297	17.252
184	18.871	17.758	17.459	17.379	17.342	18.823	17.776	17.468	17.371	17.344
185	19.349	17.966	17.496	17.279	17.223	19.349	17.948	17.533	17.292	17.257
186	18.303	16.990	16.540	16.429	16.402	18.268	17.034	16.584	16.441	16.423
187	17.711	16.545	16.173	16.084	16.070	17.664	16.540	16.195	16.095	16.064
188	19.514	18.214	17.757	17.579	17.532	19.516	18.199	17.778	17.622	17.574
189	18.636	17.554	17.250	17.147	17.127	18.660	17.577	17.278	17.178	17.146
190	18.677	17.533	17.107	16.978	16.956	18.692	17.552	17.118	16.931	16.927
191	18.593	17.397	17.043	16.914	16.873	18.630	17.433	17.045	16.910	16.869
192	19.197	18.090	17.751	17.621	17.602	19.194	18.098	17.770	17.644	17.603
193	19.209	17.369	16.637	16.376	16.211	19.208	17.369	16.662	16.383	16.216
194	18.476	17.279	16.872	16.753	16.728	18.443	17.250	16.889	16.773	16.753
195	18.915	17.908	17.567	17.426	17.416	18.903	17.873	17.578	17.453	17.465
196	19.645	18.355	17.943	17.764	17.767	19.618	18.342	17.949	17.782	17.762
197	19.058	17.711	17.300	17.173	17.118	19.080	17.736	17.312	17.203	17.122
198	19.508	17.921	17.338	17.143	17.058	19.506	17.916	17.367	17.181	17.092
199	17.685	16.498	16.025	15.889	15.867	17.693	16.524	16.059	15.917	15.883
200	19.372	17.646	16.968	16.756	16.630	19.327	17.638	16.991	16.770	16.631
201	18.804	16.937	16.298	16.142	16.060	18.791	16.942	16.312	16.141	16.066
202	19.156	18.028	17.670	17.562	17.561	19.144	18.075	17.694	17.572	17.596
203	17.689	16.624	16.236	16.128	16.073	17.697	16.614	16.241	16.124	16.101
204	18.089	16.915	16.525	16.441	16.425	18.064	16.904	16.526	16.415	16.408
205	17.474	16.397	16.082	15.969	15.947	17.485	16.415	16.084	15.963	15.954
206	18.313	17.305	17.031	16.950	16.936	18.317	17.329	17.058	16.981	16.971
207	18.536	17.433	17.075	16.967	16.903	18.501	17.439	17.091	16.969	16.931
208	18.952	17.636	17.158	17.022	16.986	18.969	17.639	17.176	17.022	16.994
209	17.816	16.602	16.171	16.092	16.043	17.789	16.609	16.182	16.100	16.056
210	18.524	17.357	16.967	16.863	16.838	18.488	17.363	16.994	16.879	16.866
211	18.911	17.634	17.179	17.017	16.990	18.947	17.677	17.223	17.063	17.023
212	20.498	17.896	16.523	15.801	15.396	20.523	17.908	16.532	15.808	15.414
213	20.765	18.720	17.895	17.630	17.472	20.722	18.718	17.907	17.643	17.488
214	17.556	16.445	16.060	15.946	15.903	17.522	16.438	16.073	15.937	15.934
215	20.964	18.664	17.227	16.404	15.973	20.934	18.648	17.246	16.408	15.974
216	19.426	17.151	16.230	15.897	15.784	19.404	17.169	16.254	15.918	15.803
217	18.691	17.362	16.905	16.747	16.737	18.656	17.361	16.923	16.741	16.711
218	18.594	17.267	16.855	16.735	16.665	18.555	17.302	16.862	16.716	16.622
219	18.395	17.041	16.552	16.393	16.343	18.442	17.069	16.577	16.436	16.367
220	20.716	18.460	17.558	17.236	17.055	20.702	18.496	17.571	17.276	17.076

221	21.135	18.564	17.336	16.868	16.600	21.116	18.550	17.339	16.905	16.633
222	18.386	17.140	16.728	16.568	16.499	18.426	17.155	16.744	16.559	16.503
223	17.917	16.886	16.608	16.516	16.510	17.944	16.920	16.631	16.542	16.545
224	19.716	18.042	17.410	17.234	17.118	19.738	18.058	17.418	17.202	17.073
225	18.475	17.240	16.849	16.720	16.683	18.497	17.251	16.854	16.731	16.678
226	19.928	18.217	17.531	17.295	17.238	19.921	18.267	17.566	17.338	17.236
227	20.828	18.570	17.594	17.230	17.054	20.837	18.608	17.602	17.244	17.015
228	18.822	17.118	16.475	16.280	16.127	18.784	17.115	16.486	16.297	16.154
229	18.381	17.027	16.674	16.540	16.547	18.375	17.077	16.707	16.575	16.579
230	19.302	17.852	17.293	17.120	17.044	19.321	17.859	17.314	17.130	17.078
231	18.949	18.069	17.794	17.714	17.695	18.974	18.035	17.798	17.703	17.682
232	19.100	17.895	17.474	17.356	17.309	19.070	17.900	17.493	17.366	17.328
233	18.310	17.244	16.892	16.774	16.729	18.346	17.271	16.910	16.795	16.777
234	18.365	16.956	16.448	16.280	16.226	18.405	16.978	16.477	16.312	16.272
235	17.582	16.686	16.426	16.327	16.323	17.600	16.720	16.471	16.371	16.366
236	20.855	18.238	16.955	16.447	16.142	20.901	18.248	16.995	16.473	16.151
237	19.872	17.872	17.092	16.832	16.677	19.883	17.920	17.121	16.848	16.683
238	18.559	17.125	16.598	16.413	16.369	18.598	17.150	16.623	16.436	16.405
239	18.224	16.967	16.641	16.452	16.409	18.199	17.003	16.656	16.477	16.430
240	18.161	16.749	16.211	16.016	15.881	18.194	16.756	16.229	16.024	15.889
241	17.832	16.507	16.036	15.896	15.840	17.864	16.549	16.063	15.918	15.856
242	19.223	18.254	17.963	17.846	17.814	19.188	18.293	17.971	17.851	17.834
243	18.130	16.924	16.457	16.316	16.277	18.124	16.958	16.468	16.321	16.277
244	19.160	17.759	17.169	17.016	16.897	19.199	17.758	17.169	16.986	16.864
245	17.615	16.594	16.251	16.151	16.155	17.631	16.613	16.270	16.170	16.184
246	18.412	16.784	16.181	15.946	15.865	18.367	16.772	16.183	15.957	15.860
247	20.569	17.776	16.381	15.232	14.640	20.541	17.789	16.412	15.248	14.643
248	18.600	17.654	17.362	17.272	17.240	18.576	17.693	17.367	17.250	17.237
249	19.287	17.682	17.133	16.944	16.881	19.290	17.719	17.159	16.965	16.896
250	20.090	17.487	16.158	15.392	14.989	20.119	17.508	16.159	15.368	14.942
251	18.665	17.304	16.827	16.692	16.627	18.688	17.321	16.832	16.689	16.627
252	18.258	16.791	16.285	16.114	16.102	18.247	16.796	16.297	16.120	16.101
253	20.285	17.662	16.332	15.757	15.459	20.242	17.677	16.345	15.774	15.473
254	21.347	18.620	17.351	15.959	15.246	21.319	18.664	17.398	15.997	15.268
255	17.853	16.614	16.153	16.027	15.954	17.892	16.659	16.182	16.035	15.975
256	18.834	17.638	17.310	17.196	17.164	18.796	17.661	17.315	17.196	17.128
257	19.790	17.765	16.934	16.667	16.516	19.774	17.781	16.943	16.673	16.498
258	18.886	17.929	17.646	17.530	17.503	18.848	17.976	17.656	17.531	17.493
259	19.101	18.101	17.634	17.533	17.403	19.144	18.097	17.655	17.562	17.413
260	19.905	17.538	16.088	14.552	13.707	19.861	17.546	16.116	14.590	13.743
261	18.209	17.420	17.100	16.974	16.981	18.246	17.444	17.116	16.991	16.995
262	20.043	17.534	16.227	14.935	14.228	20.039	17.562	16.242	14.951	14.229
263	18.569	16.749	16.097	15.916	15.869	18.584	16.745	16.111	15.924	15.860

264	19.067	18.211	17.981	17.902	17.894	19.033	18.233	18.004	17.926	17.906
265	18.304	16.809	16.174	15.924	15.858	18.325	16.807	16.187	15.928	15.845
266	18.961	17.477	16.927	16.741	16.633	18.965	17.494	16.932	16.749	16.649
267	19.742	18.205	17.723	17.468	17.350	19.755	18.215	17.724	17.494	17.367
268	18.841	17.706	17.301	17.200	17.182	18.857	17.706	17.332	17.222	17.209
269	18.782	17.608	17.151	17.000	17.004	18.749	17.612	17.169	17.035	17.004
270	19.848	17.553	16.588	16.238	16.054	19.890	17.574	16.593	16.246	16.094
271	20.010	18.228	17.530	17.320	17.216	20.008	18.258	17.565	17.335	17.203
272	20.156	17.544	16.289	15.787	15.478	20.184	17.557	16.305	15.796	15.486
273	19.600	18.229	17.732	17.572	17.513	19.553	18.242	17.757	17.592	17.537
274	20.158	17.487	16.203	15.626	15.327	20.153	17.504	16.219	15.646	15.350
275	19.778	17.854	17.080	16.803	16.643	19.781	17.842	17.102	16.845	16.678
276	17.727	16.493	16.066	15.945	15.915	17.732	16.517	16.085	15.968	15.939
277	20.111	18.366	17.717	17.462	17.361	20.077	18.401	17.758	17.484	17.405
278	18.882	17.902	17.620	17.562	17.517	18.845	17.903	17.660	17.604	17.560
279	20.270	17.613	16.316	15.785	15.507	20.264	17.628	16.332	15.795	15.535
280	17.700	16.608	16.244	16.123	16.093	17.739	16.625	16.280	16.168	16.120
281	18.143	16.954	16.501	16.337	16.311	18.128	16.951	16.502	16.337	16.327
282	19.605	18.077	17.492	17.291	17.215	19.628	18.101	17.530	17.327	17.232
283	19.599	17.981	17.429	17.205	17.113	19.617	18.010	17.453	17.204	17.097
284	18.247	17.089	16.713	16.574	16.515	18.290	17.114	16.738	16.602	16.552
285	18.015	16.723	16.207	15.948	15.856	18.020	16.697	16.210	15.981	15.896
286	19.784	18.570	18.051	17.900	17.856	19.797	18.602	18.080	17.945	17.901
287	20.561	17.816	16.554	16.092	15.807	20.526	17.847	16.577	16.115	15.821
288	19.344	17.856	17.249	17.023	16.930	19.347	17.841	17.280	17.051	16.933
289	20.223	17.747	16.382	15.699	15.344	20.209	17.765	16.407	15.724	15.368
290	18.615	17.081	16.487	16.273	16.177	18.616	17.099	16.501	16.286	16.208
291	18.167	16.996	16.611	16.485	16.410	18.202	17.032	16.627	16.508	16.440
292	19.042	18.186	17.935	17.881	17.829	19.056	18.221	17.959	17.886	17.825
293	18.878	17.916	17.615	17.528	17.469	18.894	17.951	17.650	17.534	17.487
294	18.851	17.805	17.435	17.334	17.298	18.848	17.824	17.474	17.358	17.342
295	18.535	17.452	17.089	16.980	16.964	18.539	17.482	17.129	17.011	16.971
296	18.917	18.011	17.773	17.705	17.666	18.933	18.043	17.794	17.687	17.658
297	17.956	17.027	16.737	16.642	16.710	17.981	17.059	16.744	16.619	16.671
298	18.211	16.668	16.134	15.979	15.946	18.199	16.680	16.148	15.997	15.958
299	18.581	17.649	17.416	17.317	17.323	18.600	17.674	17.428	17.326	17.355
300	19.173	18.229	17.932	17.869	17.782	19.179	18.224	17.939	17.856	17.798
301	18.274	17.162	16.739	16.631	16.555	18.286	17.174	16.785	16.671	16.601
302	18.097	17.211	16.914	16.827	16.816	18.140	17.237	16.940	16.850	16.799
303	19.938	18.514	17.943	17.708	17.596	19.946	18.509	17.946	17.742	17.642
304	19.734	17.352	16.300	15.924	15.739	19.730	17.316	16.327	15.971	15.783
305	19.040	18.091	17.825	17.715	17.740	18.994	18.083	17.839	17.746	17.723
306	18.890	17.673	17.174	16.973	16.895	18.871	17.715	17.194	16.986	16.874

307	18.829	17.606	17.154	17.005	16.973	18.848	17.598	17.156	16.997	16.945
308	18.045	16.821	16.420	16.289	16.228	18.013	16.828	16.422	16.271	16.221
309	18.132	17.001	16.509	16.431	16.395	18.148	16.973	16.517	16.431	16.429
310	19.102	17.944	17.569	17.408	17.369	19.131	17.985	17.593	17.447	17.389
311	17.994	16.818	16.423	16.279	16.234	18.035	16.842	16.449	16.319	16.274
312	17.643	16.535	16.179	16.070	16.006	17.621	16.513	16.192	16.073	15.998
313	19.032	17.451	16.875	16.676	16.584	19.057	17.483	16.915	16.711	16.610
314	18.066	16.918	16.551	16.401	16.384	18.058	16.951	16.564	16.401	16.365
315	18.947	17.505	16.980	16.818	16.769	18.955	17.537	16.998	16.820	16.733
316	18.567	16.975	16.461	16.287	16.240	18.567	17.013	16.494	16.308	16.244
317	19.293	18.108	17.653	17.480	17.441	19.306	18.132	17.701	17.487	17.469
318	19.539	18.079	17.508	17.286	17.173	19.495	18.080	17.541	17.289	17.184
319	17.751	16.518	16.104	15.958	15.945	17.710	16.524	16.124	15.986	15.976
320	18.714	17.320	16.808	16.602	16.521	18.690	17.322	16.827	16.620	16.570
321	19.515	17.673	16.983	16.730	16.603	19.510	17.714	16.991	16.739	16.577
322	17.810	16.705	16.361	16.215	16.208	17.836	16.735	16.378	16.228	16.216
323	18.390	17.295	16.911	16.776	16.773	18.385	17.291	16.917	16.805	16.786
324	17.762	16.902	16.585	16.502	16.481	17.787	16.889	16.596	16.509	16.492
325	19.419	17.490	16.743	16.493	16.342	19.412	17.504	16.754	16.501	16.368
326	18.783	17.439	16.955	16.792	16.748	18.769	17.430	16.962	16.796	16.746
327	20.580	18.138	17.003	16.611	16.343	20.599	18.154	17.006	16.582	16.305
328	18.931	17.737	17.337	17.166	17.141	18.963	17.784	17.367	17.185	17.134
329	19.582	18.041	17.448	17.248	17.123	19.625	18.032	17.448	17.244	17.130
330	18.563	17.070	16.521	16.313	16.163	18.589	17.071	16.537	16.324	16.187
331	19.406	18.078	17.601	17.410	17.357	19.366	18.063	17.628	17.458	17.350
332	19.780	18.217	17.667	17.482	17.420	19.794	18.206	17.667	17.499	17.422
333	18.974	17.424	16.890	16.697	16.644	18.939	17.428	16.898	16.696	16.634
334	18.070	16.952	16.549	16.447	16.387	18.080	16.938	16.553	16.463	16.400
335	19.169	18.120	17.819	17.667	17.647	19.164	18.149	17.830	17.645	17.630
336	19.048	16.966	16.233	16.024	15.915	19.036	16.969	16.243	16.044	15.942
337	21.096	18.891	18.002	17.683	17.549	21.142	18.894	18.025	17.699	17.555
338	18.320	16.931	16.450	16.294	16.209	18.369	16.976	16.480	16.284	16.188
339	18.507	17.192	16.813	16.689	16.615	18.529	17.213	16.828	16.695	16.613
340	20.741	17.997	16.811	16.338	16.074	20.732	17.984	16.836	16.365	16.072
341	18.565	17.453	17.102	16.990	16.924	18.552	17.473	17.115	17.011	16.960
342	19.342	17.989	17.486	17.276	17.227	19.316	18.007	17.519	17.318	17.271
343	19.127	17.910	17.435	17.300	17.239	19.119	17.886	17.442	17.313	17.257
344	19.491	18.012	17.545	17.383	17.355	19.487	18.061	17.568	17.399	17.379
345	18.919	17.914	17.585	17.443	17.410	18.938	17.914	17.586	17.471	17.454
346	19.598	18.255	17.831	17.689	17.626	19.568	18.293	17.854	17.712	17.610
347	20.130	18.574	18.003	17.802	17.752	20.097	18.601	18.020	17.789	17.733
348	18.219	17.031	16.652	16.499	16.460	18.256	17.079	16.686	16.523	16.488
349	18.518	17.316	16.896	16.734	16.691	18.538	17.322	16.905	16.745	16.705

350	18.364	17.098	16.666	16.545	16.495	18.364	17.127	16.695	16.555	16.532
351	19.244	18.225	17.914	17.820	17.805	19.198	18.232	17.918	17.817	17.787
352	19.428	17.629	16.969	16.741	16.648	19.476	17.650	16.979	16.751	16.646
353	19.190	18.060	17.719	17.587	17.581	19.179	18.077	17.741	17.607	17.593
354	18.568	16.743	16.069	15.837	15.789	18.594	16.761	16.088	15.849	15.779
355	18.782	17.505	17.054	16.867	16.830	18.744	17.506	17.092	16.889	16.863
356	19.116	17.999	17.585	17.475	17.449	19.066	17.969	17.613	17.479	17.483
357	18.731	17.579	17.156	17.017	16.983	18.691	17.593	17.178	17.061	17.009
358	19.486	18.356	17.990	17.846	17.779	19.471	18.383	18.006	17.860	17.829
359	20.711	18.271	17.287	16.876	16.653	20.718	18.294	17.301	16.888	16.646
360	20.207	18.375	17.634	17.377	17.257	20.200	18.380	17.667	17.418	17.278
361	17.743	16.615	16.203	16.099	16.071	17.748	16.644	16.221	16.118	16.072
362	19.025	17.767	17.270	17.093	16.995	18.986	17.750	17.288	17.112	17.039
363	18.951	17.740	17.378	17.224	17.198	18.977	17.765	17.406	17.265	17.230
364	18.461	17.307	16.923	16.804	16.752	18.475	17.326	16.936	16.794	16.753
365	18.665	17.395	16.983	16.845	16.817	18.648	17.394	16.986	16.850	16.822
366	19.316	18.185	17.790	17.638	17.631	19.346	18.188	17.798	17.614	17.598
367	20.846	18.634	17.567	17.217	17.015	20.825	18.622	17.570	17.216	16.989
368	21.064	18.586	17.450	17.058	16.817	21.092	18.593	17.485	17.076	16.847
369	17.980	16.538	16.079	15.936	15.900	18.029	16.562	16.086	15.930	15.910
370	18.125	16.885	16.415	16.278	16.210	18.098	16.905	16.459	16.324	16.245
371	19.200	18.317	18.040	17.995	17.913	19.229	18.353	18.057	17.989	17.961
372	18.369	17.250	16.893	16.787	16.741	18.367	17.259	16.908	16.814	16.780
373	17.890	16.807	16.473	16.359	16.340	17.911	16.855	16.521	16.388	16.348
374	18.494	17.314	16.936	16.725	16.794	18.504	17.347	16.939	16.748	16.779
375	18.705	17.559	17.139	17.049	16.984	18.665	17.553	17.141	17.047	16.981
376	19.343	18.005	17.427	17.232	17.127	19.373	18.020	17.439	17.224	17.174
377	19.347	18.114	17.766	17.641	17.644	19.391	18.157	17.775	17.656	17.672
378	17.956	16.800	16.362	16.249	16.156	17.983	16.832	16.378	16.267	16.173
379	19.522	18.257	17.775	17.662	17.577	19.568	18.301	17.805	17.625	17.591
380	19.563	17.693	16.990	16.792	16.676	19.573	17.663	16.991	16.792	16.669
381	18.357	17.082	16.645	16.523	16.490	18.378	17.086	16.649	16.529	16.527
382	19.564	18.134	17.599	17.410	17.350	19.521	18.162	17.629	17.419	17.339
383	19.332	18.172	17.694	17.549	17.466	19.312	18.165	17.703	17.559	17.507
384	18.650	17.255	16.799	16.616	16.557	18.629	17.269	16.805	16.612	16.524
385	19.904	17.551	16.582	16.261	16.120	19.916	17.533	16.591	16.254	16.102
386	18.007	17.195	16.896	16.789	16.759	18.040	17.226	16.904	16.774	16.750
387	18.408	17.250	16.798	16.645	16.555	18.404	17.256	16.825	16.689	16.603
388	18.398	17.450	17.091	16.942	16.936	18.417	17.428	17.096	16.986	16.959

#	H	M	S	D	M	S	MAG 1	MAG2	SEP	PA
389	7	51	35.07	15	42	20.69	19.15	19.174	29.973	287.667
390	7	53	17.19	8	30	41.69	19.083	19.087	27.125	325.025
391	7	54	56.5	-1	46	35.58	19.599	19.635	29.537	292.743
392	7	55	44.87	9	8	21.09	19.26	19.294	20.998	302.58
393	7	56	6.59	8	41	0.42	19.068	19.087	19.532	348.681
394	7	56	42.17	14	34	9.24	18.496	18.499	29.21	20.803
395	7	57	17.81	8	48	11.12	18.126	18.155	17.502	227.231
396	8	0	19.2	5	57	19.09	18.172	18.174	21.71	175.443
397	8	1	34.41	16	55	40.33	18.877	18.883	14.292	334.878
398	8	3	49.24	3	31	44.15	18.399	18.406	21.54	227.597
399	8	8	18.16	0	38	14.35	18.265	18.267	18.034	75.122
400	8	8	23.74	4	46	11.88	18.369	18.387	9.7	148.021
401	8	13	58.02	10	3	31.71	18.354	18.384	26.685	243.141
402	8	20	5.09	12	40	25.8	18.648	18.653	6.463	275.976
403	8	22	6.58	5	35	10.7	18.121	18.148	3.449	24.487
404	8	26	31.64	4	23	31.46	18.051	18.061	14.983	50.725
405	8	33	53.55	41	59	40.46	19.127	19.14	23.265	219.532
406	8	37	46	17	37	37.21	18.16	18.207	28.843	174.787
407	9	2	14.75	43	25	31.59	18.078	18.127	12.351	181.356
408	9	31	57	24	30	29.7	19.434	19.482	13.859	60.012
409	9	35	29.89	52	2	19.03	19.572	19.58	23.856	15.476
410	10	25	52.77	4	38	36.45	19.287	19.317	27.81	16.412
411	10	34	41.03	8	35	27.74	19.032	19.051	28.346	223.192
412	10	50	56.93	7	55	40.14	18.283	18.308	18.919	340.995
413	10	58	52.95	28	42	44.32	18.281	18.309	21.204	276.968
414	11	0	26.15	47	42	36.4	19.352	19.365	20.128	204.969
415	11	6	24.27	21	59	32.67	19.643	19.646	28.265	136.111
416	11	32	18.66	36	2	28.25	18.581	18.583	19.867	207.895
417	12	5	17.27	1	55	50	19.551	19.552	23.84	317.485
418	12	6	5.22	24	3	1.09	19.588	19.6	13.585	70.298
419	12	14	43.49	49	47	35.13	19.872	19.874	19.51	205.452
420	12	16	3.59	15	43	17.25	19.068	19.088	29.375	117.466
421	12	31	41.57	-3	37	9.69	19.24	19.255	21.434	166.164
422	12	56	22.4	30	49	22.7	18.638	18.667	23.434	130.716
423	13	8	58.47	58	2	22.68	19.445	19.465	16.347	221.769
424	13	16	12.59	17	38	37.08	19.401	19.429	13.92	72.318
425	13	16	16.68	17	38	20.42	19.566	19.59	14.778	57.579
426	13	17	18.77	27	4	29.91	18.702	18.715	24.175	84.839
427	13	28	45.4	36	47	31.03	19.12	19.145	29.861	21.607
428	13	32	48.53	23	6	35.62	18.948	18.982	22.137	144.519

429	13	34	1.75	17	2	38.28	19.761	19.773	29.995	35.589
430	13	36	5.41	8	28	28.9	18.859	18.875	14.917	170.369
431	13	41	50.94	28	9	29.96	18.483	18.5	15.755	293.987
432	13	42	55.64	4	37	13.4	18.839	18.889	4.869	164.075
433	13	43	7.62	0	5	8.79	19.486	19.493	13.906	178.417
434	13	47	39.87	45	46	7.54	19.724	19.763	23.915	18.742
435	13	53	35.82	10	33	25.81	19.202	19.239	23.881	283.52
436	13	54	1.97	19	21	6.2	19.853	19.902	25.573	355.009
437	13	55	42.36	-6	19	39.21	18.287	18.315	15.839	27.508
438	14	21	51.53	23	45	16.4	19.501	19.501	6.863	251.185
439	14	22	35.46	20	25	11.76	19.558	19.568	27.149	306.7
440	14	27	38	9	19	8.14	18.068	18.083	24.149	323.932
441	14	37	55.43	0	36	29.3	19.236	19.25	26.314	320.996
442	14	50	22.39	6	6	1.79	18.388	18.393	24.964	78.956
443	14	55	11.67	14	6	5.91	18.347	18.353	15.685	105.729
444	15	13	19.76	27	50	58.5	18.783	18.796	17.447	334.609
445	15	24	31.67	32	12	6.17	19.273	19.283	21.344	22.757
446	15	27	34.09	45	40	29.82	20.031	20.037	26.582	301.142
447	15	45	54.15	16	48	17.36	18.824	18.855	28.351	301.212
448	16	2	28.42	45	46	27.51	19.386	19.41	17.87	39.629
449	16	5	8.83	27	50	17.91	19.035	19.056	19.954	256.359
450	16	6	51.09	32	19	47.68	18.039	18.05	13.337	285.053
451	16	27	1.61	28	51	59.17	19.589	19.637	20.641	35.345
452	16	33	32.6	25	47	43.07	19.752	19.798	5.885	129.082
453	16	35	8.91	55	27	37.14	18.772	18.788	25.303	76.806
454	16	41	31.11	36	36	57.49	18.663	18.688	20.773	355.427
455	16	41	59.26	36	19	17.1	19.488	19.499	20.262	341.26
456	16	41	59.76	36	18	23.88	19.564	19.586	15.63	183.874
457	16	42	1.97	36	39	5.37	19.211	19.223	27.295	250.378
458	16	42	3.28	36	17	52.78	19.61	19.623	28.067	137.514
459	16	42	3.74	36	18	46.41	18.242	18.267	29.465	359.076
460	16	42	4.19	35	27	1.54	18.365	18.413	17.407	188.245
461	16	42	28.62	29	46	38.88	19.347	19.364	27.914	325.555
462	16	48	9.03	32	2	48.82	18.674	18.72	23.007	214.799
463	16	58	14.38	32	31	56.78	19.427	19.428	27.881	160.158
464	17	0	0.82	34	39	59.17	18.333	18.351	9.097	108.007
465	17	2	29.41	35	24	45.51	19.814	19.819	14.861	228.29
466	17	3	47.37	61	22	6.18	19.323	19.327	14.759	10.434
467	17	4	54.6	36	34	20.03	18.152	18.164	10.944	33.727
468	17	7	0.47	33	28	1.48	18.79	18.84	21.634	178.001
469	17	8	14.26	31	41	46.26	19.784	19.794	20.848	55.253
470	17	14	24.03	37	10	6.95	19.005	19.043	28.273	258.817
471	17	14	32.88	59	17	17.73	19.471	19.48	10.306	27.369

472	17	17	34.64	43	0	18.72	19.443	19.474	25.161	154.175
473	17	19	52.45	57	57	28.08	19.327	19.35	25.727	226.428
474	17	24	18.66	61	26	51.34	19.56	19.588	21.845	341.18
475	17	36	48.54	43	7	56.54	19.239	19.268	15.686	214.839
476	17	39	36.51	63	46	16.95	19.637	19.649	6.85	198.792
477	17	40	10.91	50	40	0.17	18.481	18.519	26.605	141.164
478	17	45	15.38	44	50	12.78	18.983	18.998	20.46	154.477
479	18	3	33.53	47	21	24.24	18.166	18.171	19.819	16.668
480	20	36	54.14	-6	51	54.76	19.406	19.434	28.775	109.064
481	21	30	51.63	-6	54	47.16	18.786	18.792	13.059	105.481

Photometry for Batch 3

	Prim u	Prim g	Prim r	Prim i	Prim z	Sec u	Sec g	Sec r	Sec i	Sec z
389	20.692	19.578	19.150	19.100	18.998	20.694	19.569	19.174	19.052	19.006
390	20.821	19.565	19.083	18.947	18.933	20.871	19.578	19.087	18.991	18.905
391	20.876	20.024	19.599	19.510	19.485	20.892	20.029	19.635	19.546	19.520
392	20.572	19.578	19.260	19.203	19.194	20.605	19.569	19.294	19.220	19.233
393	21.068	19.602	19.068	18.913	18.875	21.042	19.600	19.087	18.922	18.839
394	20.248	18.999	18.496	18.374	18.306	20.279	19.002	18.499	18.335	18.312
395	19.920	18.602	18.126	17.950	17.924	19.891	18.641	18.155	17.979	17.963
396	19.457	18.425	18.172	18.083	18.091	19.484	18.442	18.174	18.064	18.053
397	20.229	19.186	18.877	18.795	18.744	20.182	19.208	18.883	18.804	18.756
398	19.648	18.697	18.399	18.271	18.233	19.628	18.722	18.406	18.274	18.282
399	19.701	18.608	18.265	18.086	18.107	19.651	18.569	18.267	18.097	18.091
400	19.555	18.591	18.369	18.350	18.341	19.599	18.636	18.387	18.343	18.378
401	19.752	18.734	18.354	18.229	18.234	19.750	18.699	18.384	18.270	18.252
402	20.799	19.228	18.648	18.434	18.331	20.762	19.235	18.653	18.421	18.291
403	19.362	18.453	18.121	18.030	18.032	19.367	18.449	18.148	18.069	18.056
404	19.413	18.386	18.051	17.953	17.899	19.384	18.374	18.061	17.955	17.934
405	20.567	19.373	19.127	19.098	19.014	20.606	19.405	19.140	19.082	18.982
406	19.834	18.602	18.160	18.056	17.967	19.848	18.617	18.207	18.040	17.976
407	19.630	18.449	18.078	17.889	17.924	19.677	18.484	18.127	17.892	17.906
408	20.820	19.877	19.434	19.303	19.285	20.776	19.859	19.482	19.341	19.309
409	20.755	19.863	19.572	19.458	19.333	20.769	19.912	19.580	19.448	19.376
410	20.741	19.663	19.287	19.165	19.095	20.739	19.700	19.317	19.165	19.137
411	20.142	19.281	19.032	18.942	18.951	20.135	19.307	19.051	18.942	18.960
412	19.504	18.556	18.283	18.198	18.155	19.530	18.588	18.308	18.177	18.126
413	19.445	18.507	18.281	18.152	18.136	19.444	18.534	18.309	18.167	18.175
414	20.699	19.756	19.352	19.229	19.101	20.699	19.720	19.365	19.262	19.147
415	21.100	20.094	19.643	19.502	19.396	21.087	20.068	19.646	19.474	19.368
416	19.722	18.862	18.581	18.497	18.503	19.763	18.889	18.583	18.475	18.465
417	20.639	19.804	19.551	19.540	19.384	20.593	19.782	19.552	19.502	19.375
418	20.884	19.880	19.588	19.480	19.468	20.846	19.918	19.600	19.477	19.472
419	21.055	20.195	19.872	19.742	19.812	21.028	20.180	19.874	19.747	19.789
420	20.188	19.356	19.068	18.957	18.980	20.150	19.325	19.088	18.984	18.952
421	21.071	19.745	19.240	19.037	18.936	21.082	19.746	19.255	19.063	18.973
422	19.884	18.933	18.638	18.530	18.520	19.837	18.942	18.667	18.576	18.558
423	20.576	19.730	19.445	19.306	19.304	20.626	19.757	19.465	19.338	19.335
424	20.527	19.613	19.401	19.281	19.224	20.505	19.633	19.429	19.320	19.217
425	20.613	19.720	19.566	19.500	19.501	20.598	19.685	19.590	19.465	19.460
426	19.912	18.951	18.702	18.620	18.666	19.911	18.906	18.715	18.663	18.649
427	20.342	19.406	19.120	19.042	19.043	20.338	19.444	19.145	19.040	19.008
428	20.170	19.273	18.948	18.810	18.733	20.179	19.289	18.982	18.820	18.759

429	20.802	20.015	19.761	19.755	19.846	20.813	20.041	19.773	19.710	19.802
430	20.094	19.116	18.859	18.749	18.746	20.087	19.146	18.875	18.746	18.735
431	19.650	18.733	18.483	18.428	18.447	19.656	18.731	18.500	18.466	18.476
432	20.416	19.325	18.839	18.739	18.664	20.454	19.347	18.889	18.766	18.665
433	20.769	19.833	19.486	19.403	19.360	20.730	19.814	19.493	19.444	19.332
434	20.874	20.017	19.724	19.679	19.675	20.827	20.043	19.763	19.702	19.651
435	20.562	19.581	19.202	19.087	19.060	20.567	19.615	19.239	19.079	19.033
436	21.032	20.135	19.853	19.812	19.716	21.068	20.181	19.902	19.772	19.744
437	19.828	18.729	18.287	18.179	18.065	19.834	18.777	18.315	18.215	18.082
438	20.721	19.792	19.501	19.429	19.347	20.711	19.828	19.501	19.432	19.361
439	20.862	19.931	19.558	19.379	19.365	20.835	19.950	19.568	19.417	19.337
440	21.194	18.997	18.068	17.747	17.547	21.188	18.976	18.083	17.779	17.578
441	20.428	19.481	19.236	19.160	19.145	20.402	19.474	19.250	19.181	19.140
442	20.017	18.846	18.388	18.208	18.149	20.066	18.847	18.393	18.244	18.141
443	20.033	18.848	18.347	18.184	18.124	20.045	18.823	18.353	18.189	18.136
444	20.087	19.097	18.783	18.686	18.680	20.093	19.134	18.796	18.703	18.718
445	21.252	19.855	19.273	19.096	18.929	21.234	19.846	19.283	19.093	18.902
446	21.265	20.330	20.031	19.988	19.874	21.264	20.296	20.037	19.969	19.844
447	20.316	19.245	18.824	18.679	18.585	20.304	19.241	18.855	18.707	18.613
448	20.856	19.823	19.386	19.221	19.288	20.902	19.779	19.410	19.262	19.253
449	20.379	19.349	19.035	18.945	18.932	20.330	19.324	19.056	18.949	18.923
450	19.085	18.257	18.039	17.934	17.938	19.112	18.262	18.050	17.925	17.932
451	20.861	19.895	19.589	19.514	19.511	20.843	19.883	19.637	19.553	19.559
452	20.919	20.052	19.752	19.714	19.678	20.938	20.086	19.798	19.702	19.700
453	20.251	19.157	18.772	18.615	18.553	20.249	19.185	18.788	18.617	18.599
454	19.807	18.856	18.663	18.571	18.602	19.844	18.901	18.688	18.611	18.611
455	20.730	19.803	19.488	19.301	19.324	20.739	19.794	19.499	19.296	19.373
456	20.771	19.904	19.564	19.429	19.497	20.798	19.899	19.586	19.438	19.447
457	20.274	19.465	19.211	19.102	19.074	20.302	19.476	19.223	19.076	19.033
458	20.975	19.868	19.610	19.472	19.437	20.927	19.902	19.623	19.439	19.408
459	19.360	18.463	18.242	18.115	18.196	19.384	18.508	18.267	18.145	18.240
460	19.584	18.661	18.365	18.313	18.290	19.569	18.689	18.413	18.362	18.338
461	20.669	19.597	19.347	19.231	19.275	20.652	19.601	19.364	19.233	19.250
462	21.176	19.423	18.674	18.401	18.268	21.134	19.413	18.720	18.422	18.261
463	20.569	19.633	19.427	19.314	19.313	20.618	19.635	19.428	19.296	19.330
464	19.632	18.644	18.333	18.224	18.211	19.639	18.686	18.351	18.252	18.203
465	20.852	20.074	19.814	19.674	19.806	20.869	20.075	19.819	19.714	19.758
466	21.131	19.820	19.323	19.217	19.104	21.154	19.835	19.327	19.169	19.128
467	20.396	18.800	18.152	17.900	17.805	20.390	18.787	18.164	17.926	17.798
468	20.644	19.317	18.790	18.635	18.553	20.692	19.366	18.840	18.655	18.545
469	21.048	20.142	19.784	19.697	19.647	21.048	20.138	19.794	19.700	19.637
470	20.526	19.447	19.005	18.858	18.845	20.526	19.484	19.043	18.834	18.799
471	20.475	19.713	19.471	19.334	19.402	20.479	19.721	19.480	19.339	19.445

472	20.444	19.695	19.443	19.369	19.298	20.475	19.674	19.474	19.364	19.277
473	21.171	19.895	19.327	19.054	18.935	21.142	19.892	19.350	19.048	18.968
474	20.853	19.807	19.560	19.471	19.495	20.852	19.804	19.588	19.520	19.482
475	21.148	19.765	19.239	19.069	18.955	21.144	19.779	19.268	19.082	18.967
476	21.025	19.957	19.637	19.533	19.552	20.988	19.943	19.649	19.546	19.530
477	21.088	19.248	18.481	18.217	18.111	21.044	19.247	18.519	18.238	18.125
478	20.786	19.505	18.983	18.816	18.728	20.775	19.468	18.998	18.832	18.771
479	19.623	18.468	18.166	17.969	17.930	19.600	18.479	18.171	17.995	17.930
480	20.692	19.754	19.406	19.289	19.178	20.658	19.746	19.434	19.279	19.200
481	20.384	19.179	18.786	18.627	18.584	20.358	19.228	18.792	18.638	18.547

Acknowledgements:

This research has made use of the VizieR catalogue access tool, CDS, Strasbourg, France.

Funding for SDSS-III has been provided by the Alfred P. Sloan Foundation, the Participating Institutions, the National Science Foundation, and the U.S. Department of Energy Office of Science. The SDSS-III web site is http://www.sdss3.org/.

SDSS-III is managed by the Astrophysical Research Consortium for the Participating Institutions of the SDSS-III Collaboration including the University of Arizona, the Brazilian Participation Group, Brookhaven National Laboratory, University of Cambridge, Carnegie Mellon University, University of Florida, the French Participation Group, the German Participation Group, Harvard University, the Instituto de Astrofisica de Canarias, the Michigan State/Notre Dame/JINA Participation Group, Johns Hopkins University, Lawrence Berkeley National Laboratory, Max Planck Institute for Astrophysics, Max Planck Institute for Extraterrestrial Physics, New Mexico State University, New York University, Ohio State University, Pennsylvania State University, University of Portsmouth, Princeton University, the Spanish Participation Group, University of Tokyo, University of Utah, Vanderbilt University, University of Virginia, University of Washington, and Yale University.

References:

Nicholson, M. (2009). A Critique of a method for Identifying Common Proper-Motion Pairs. *PHILICA.COM Observation number 55.*

BY THE SAME AUTHOR

All are available from Amazon.com and from Amazon.co.uk

1800 new double stars for amateur observers

3600 celestial asterisms for amateur astronomers

Discover your own double star

Discover your own variable star

Identifying Common Proper Motion Binary Star Systems

Identifying Identical Twin Star Systems from the SDSS Data Release 10

www.ingramcontent.com/pod-product-compliance
Lightning Source LLC
Chambersburg PA
CBHW081809170526
45167CB00008B/3386